Planning Ethical Influence Operations

A Framework for Defense Information Professionals

CHRISTOPHER PAUL, WILLIAM MARCELLINO, MICHAEL SKERKER,
JEREMY DAVIS, BRADLEY J. STRAWSER

T0321469

Prepared for the Office of the Secretary of Defense
Approved for public release; distribution is unlimited

RAND NATIONAL DEFENSE RESEARCH INSTITUTE

For more information on this publication, visit **www.rand.org/t/RRA1969-1**.

About RAND

The RAND Corporation is a research organization that develops solutions to public policy challenges to help make communities throughout the world safer and more secure, healthier and more prosperous. RAND is nonprofit, nonpartisan, and committed to the public interest. To learn more about RAND, visit www.rand.org.

Research Integrity

Our mission to help improve policy and decisionmaking through research and analysis is enabled through our core values of quality and objectivity and our unwavering commitment to the highest level of integrity and ethical behavior. To help ensure our research and analysis are rigorous, objective, and nonpartisan, we subject our research publications to a robust and exacting quality-assurance process; avoid both the appearance and reality of financial and other conflicts of interest through staff training, project screening, and a policy of mandatory disclosure; and pursue transparency in our research engagements through our commitment to the open publication of our research findings and recommendations, disclosure of the source of funding of published research, and policies to ensure intellectual independence. For more information, visit www.rand.org/about/principles.

RAND's publications do not necessarily reflect the opinions of its research clients and sponsors.

Published by the RAND Corporation, Santa Monica, Calif.
© 2023 RAND Corporation
RAND® is a registered trademark.

Library of Congress Cataloging-in-Publication Data is available for this publication.

ISBN: 978-1-9774-1197-6

Cover: U.S. Air Force photo by Tech. Sgt. Devin Nothstine, and kertlis/Getty Images.

About This Report

Many Americans view influence pejoratively, equating it with manipulation, disinformation, or propaganda. How should the U.S. Department of Defense ethically plan and conduct influence operations? In this report, we explore how ethical considerations are raised within the influence planning and approval process, review relevant thinking and scholarship in the field of applied ethical philosophy, and present a framework for making ethical judgments and justifications about proposed influence efforts that is intended to be useful for both planners and approvers.

The research reported here was completed in March 2023 and underwent security review with the sponsor and the Defense Office of Prepublication and Security Review before public release.

RAND National Security Research Division

This research was conducted within the International Security and Defense Policy Program of the RAND National Security Research Division (NSRD), which operates the RAND National Defense Research Institute (NDRI), a federally funded research and development center (FFRDC) sponsored by the Office of the Secretary of Defense, the Joint Staff, the Unified Combatant Commands, the Navy, the Marine Corps, the defense agencies, and the defense intelligence enterprise. This research was made possible by NDRI exploratory research funding that was provided through the FFRDC contract and approved by NDRI's primary sponsor.

For more information on the RAND International Security and Defense Policy Program, see www.rand.org/nsrd/isdp or contact the director (contact information is provided on the webpage).

Acknowledgments

We are deeply indebted to the current and former defense influence practitioners and approval process participants who spoke with us as part of this effort. The terms of our interviews preclude us from thanking you by name, but you know who you are, and we know that your insights and observations were critical to shaping this report. We also thank Maria Falvo for support with formatting and administrative support related to our collaboration across institutions. We truly appreciate the feedback provided by Todd Helmus and Ryan Jenkins, and we thank the RAND publications team: Daphne Rozenblatt, Allison Kerns, and Matthew Byrd.

Summary

Many Americans view influence pejoratively, equating it with manipulation, disinformation, or propaganda. Nevertheless, the U.S. Department of Defense (DoD) has influence tasks, capabilities, and missions. This report considers the following questions: Are influence operations ethical? What factors, conditions, or considerations affect the rightness of such actions? Does the target, location, content, or veracity of the communication make a difference? How should DoD ethically plan and conduct influence operations?

To answer these questions, we offer a principles-based framework for military practitioners to determine whether a proposed influence effort is ethically permissible. We also offer guidance for the preparation of a justification statement that would allow approvers to follow the ethical logic behind a proposed influence effort and either agree or disagree with the practitioner's determination.

The Challenge

DoD efforts to plan and conduct influence operations in an ethical manner face several challenges. These challenges include (1) general and broad concerns regarding the appropriateness of any form of influence activity because of a cultural distaste for manipulation, which draws on certain historical episodes in which influence efforts raised serious ethical questions, (2) a lack of explicit consideration of ethics in the influence-planning process (with a false equivalency often made between *legal* and *ethical*, as well as approvers having ethical qualms but lacking the vocabulary to describe them), and (3) decoupling the ethics of force from the ethics of influence in military operations.

Currently, DoD lacks influence-specific ethics frameworks, and there is no formal explicit consideration of ethics in the influence activity approval process aside from legal review. Among the subject-matter experts we interviewed, there was a strong consensus that this is an important gap.

Insights from Ethics Scholarship

Reviewing the ethics literature and scholarship, we found that the principal ethical objection to influence is its threat to autonomy. A person's right to determine their preferences and actions is among the most fundamental of rights. Influence activities that involve deception or manipulation violate autonomy because the target would presumably make a different choice if they knew the true facts or the true source of the information (leaving someone an independent choice by persuading them with temperate and true facts is not a violation of their autonomy). Similarly, emotionally manipulative influence seeks to bypass a target's reason (either their logical reasoning or their emotional reasoning) and is thus also a threat to autonomy. Threats and coercion are also violations of autonomy. Violations of autonomy are a form of harm, and, in most situations, it is morally wrong to harm others. Doing so would require a justification from DoD.

Although influence is a threat to autonomy and thus morally fraught, existing scholarship points to several situations in which influence activities might be justified. The first is *virtuous persuasion*, which is reasoned argumentation that does not threaten autonomy. Virtuous persuasion is always permissible. Also, it is permissible to manipulate or deceive those who are otherwise liable to harm. For example, enemy combatants who are legitimately subject to lethal force under the rules of engagement can be understood to be *liable* to physical harm, up to and including death. By logical extension, they are also liable to the lesser harms of deception and manipulation. It is harder to justify deceiving or manipulating targets who are not enemy combatants and therefore not liable to (and should be protected from) physical harm. The principles of just war theory can provide guidance for determining whether influencing noncombatants can be justified. These principles include just cause, proportionality, last resort or necessity, and discrimination.

Practical Application

We propose that practitioners follow three principles when considering ethics in influence planning: *necessity*, *effectiveness*, and *proportionality*. These principles are implemented in the framework proposed here through five criteria: Military influence efforts should (1) seek *legitimate military outcomes*, (2) be *necessary* to attain those outcomes, (3) employ *means that are not harmful* (or harm only those liable to harm), (4) have *high likelihoods of success*, and (5) should *not generate second-order effects* beyond what is intended. Efforts that do not wholly satisfy all criteria might still be justified, but only if the harm likely to be caused is substantially outweighed by the expected benefit. The framework structuring the application of these principles is intended to unfold in three phases: an initial screening, followed (if necessary) by a full ethical risk assessment, and finally the preparation of a justification statement. The workflow proposed by the framework is captured in Figure S.1.

Note that this framework recognizes that different individuals might reach different conclusions regarding the permissibility of certain proposed actions. That is expected and perfectly reasonable. This framework and approach promote consensus on process rather than outcome: All planners, practitioners, and reviewers

FIGURE S.1
Workflow of the Application of the Framework for Ethical Influence

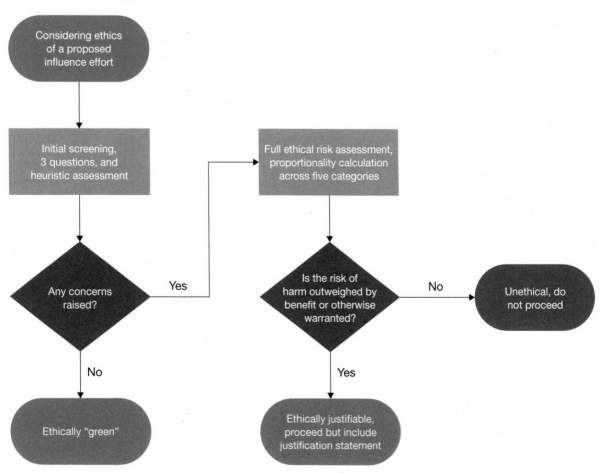

should agree that it is appropriate and even morally necessary to make an ethical assessment of proposed influence operations; unless some very strict criteria are unambiguously satisfied, it is necessary and appropriate to write a justification statement related to the ethics of a proposed effort; and a common vocabulary that describes what is and is not acceptable and why can help make points of disagreement clear and support movement toward resolution and consensus for individual proposals.

Recommendations

Going forward, joint and service doctrine relevant to influence operations should include the consideration of ethical concerns. Planners and practitioners should adhere to the ethical principles presented here and should follow a logical process to make ethical determinations, either using the process detailed here or another that bears similar results. An ethical review process like this one is currently missing from the review and approval process for influence operations. It or something like it should be formally added to these processes to provide greater clarity for planners and reviewers and increase process discipline.

Contents

Figures

CHAPTER 1

Introduction

Many Americans view influence pejoratively, equating it with manipulation, disinformation, or propaganda. For example, there was public consternation and recrimination following an August 2022 report from the Stanford Internet Observatory describing an interconnected web of accounts on eight social media platforms engaged in coordinated, inauthentic behavior to advance narratives favorable to the United States and its allies. The report suggested that these accounts were run by the U.S. government or presumptively originated from the United States.[1]

If this did occur, was it ethical? What factors, conditions, or considerations affect the rightness of such actions? Does the target, location, content, or veracity of the communication make a difference? How should the U.S. Department of Defense (DoD) ethically plan and conduct influence operations?

This report provides answers to these questions and a framework for military practitioners to use to help determine whether a proposed influence effort is ethically permissible. As one practitioner of influence operations we interviewed put it, "Senior leaders think this an intuitive process, and it is NOT. You need a cookbook, a rulebook, a set of guidelines."[2] This need for clear guidelines is the motivation for this report. The framework we propose will help practitioners make ethical determinations and produce justification statements that will allow approvers to follow the ethical logic behind a proposed influence effort and either agree or disagree.

What Do We Mean by Influence Operations?

The term *influence operations* is not included in the DoD lexicon.[3] However, JP 3-04, *Information in Joint Operations*, alludes to such operations when it defines *operations in the information environment* (OIE) as "[m]ilitary actions involving the integrated employment of multiple information forces to affect drivers of behavior."[4] This definition implies that affecting the drivers of human behavior is synonymous with influence. JP 3-04 lays out several tasks related to information for the joint force, one of which is "leverage information," which includes a subtask to "influence foreign relevant actors."[5] Furthermore, the doctrine notes

---

[1] Graphika and Stanford Internet Observatory, *Unheard Voice: Evaluating Five Years of Pro-Western Covert Influence Operations*, Stanford Digital Repository, August 24, 2022.

[2] DoD civilian employee with oversight responsibilities related to influence, interview with the authors, August 18, 2022.

[3] Note the acronym for influence operations might be IO. However, IO is shorthand for *information operations*, a frequently misunderstood term that was removed from the joint lexicon by the publication of Joint Publication (JP) 3-04, *Information in Joint Operations*, Joint Chiefs of Staff, September 14, 2022. We avoid the term (and the acronym) to avoid any further confusion. For a discussion of the merits of the term, see Christopher Paul, "Is It Time to Abandon the Term Information Operations?" *Strategy Bridge*, March 11, 2019.

[4] JP 3-04, 2022, p. GL-5.

[5] JP 3-04, 2022, pp. viii, x.

that the subtask might involve the deliberate selection of capabilities based on their inherent informational aspects, including the adjustment of the timing, duration, or visibility of an operation but also the employment of a designated force to conduct OIE or the employment of individual information forces. Therefore, we define *influence operations* as the use of information forces to contribute to the "influence foreign relevant actors" subtask.

We narrow the scope of influence operations to the efforts of information forces. In other words, we exclude efforts that primarily leverage the inherent informational aspects of other military activities, such as the implied threat of a tank or carrier strike group's presence or the influence of physical destruction on individuals. For example, the destruction of a bridge influences anyone who wishes to get to the other side, and killing someone prevents them from undertaking any undesired behavior. Arguably, all warfare is a form of influence, but the ethics governing the use of physical force in war are more fully understood than the ethics related to influence through the use of information.[6]

The Ethical Challenge Posed by Influence Operations

DoD efforts to plan and conduct influence operations in an ethical manner face several challenges. These include general and broad concerns regarding the appropriateness of any form of influence activity because of a significant cultural distaste for manipulation, certain historical episodes in which influence efforts raised serious ethical questions, a lack of explicit consideration of ethics in the influence-planning process (with a false equivalency often made between *legal* and *ethical* and approvers having ethical qualms but lacking vocabulary to describe them), and decoupling the ethics of force from the ethics of influence in military operations. Each of these challenges is described in greater detail below.

Americans generally view government *propaganda* pejoratively and express skepticism about the ethical appropriateness of any kind of government or military influence effort, especially if it is in any way misleading or deceptive.[7] Some historical U.S. military influence operations have exacerbated this distrust of influence because they have been unethical; in turn, this has increased fears of future or ongoing unethical influence and increased skepticism about all influence operations.

While deception is a part of military practice with ancient roots (consider, for example, the Trojan Horse or Sun Tzu's maxim that "all warfare is based on deception"[8]), deception based on false public statements remain ethically dubious. Deception by maneuver is generally considered to be acceptable; it creates a false impression for an enemy based on troop positions or movement (such as the famous "left hook" of Operation Desert Storm). Other forms of deception are viewed more critically. For example, Marine Lt. Lyle Gilbert's October 14, 2004, announcement that troops had crossed the line of departure to begin operations in Fallujah, Iraq, when combat operations would not commence there for another three weeks was met with understandable opprobrium.[9] Sometimes, physical deceptions can raise ethical issues as well. Deception designs

---

[6] For more on war as a form of influence, see Scott K. Thomson and Christopher E. Paul, "Paradigm Change: Operational Art and the Information Joint Function," *Joint Force Quarterly*, Vol. 89, 2nd Quarter 2018. For discussions regarding the ethics of the use of force in war, see David Rodin, *War and Self-Defense*, Oxford University Press, 2002; Ryan Jenkins, Michael Robillard, and Bradley Jay Strawser, eds., *Who Should Die? The Ethics of Killing in War*, Oxford University Press, 2018; Helen Frowe, *Defensive Killing: An Essay on War and Self-Defence*, Oxford University Press, 2014; and Jeff McMahan, *Killing in War*, Oxford University Press, 2009.

[7] For a discussion on propaganda, see Christopher Paul, *Information Operations—Doctrine and Practice: A Reference Handbook*, Praeger Security International, 2008, pp. 8–9.

[8] Sun Tzu, *The Art of War*, trans. by Samuel B. Griffith, Oxford University Press, 1963, p. 106.

[9] Mark Mazzetti, "PR Meets Psy-Ops in War on Terror," *Los Angeles Times*, December 1, 2004.

that are strange and perhaps somewhat nonsensical can contribute to public skepticism (such as the parachuting of frozen blocks of animal blood over the North Vietnamese jungle, which were meant to distract North Vietnamese forces with the prospect of finding a wounded American pilot).[10] The U.S. public can even become incensed when government organization names or missions sound like they might inappropriately use information or influence. Consider the ignominious 2002 demise of DoD's Office of Strategic Influence after the office was suspected of intending to plant fabricated news items in international media.[11] Another example is the backlash that forced the U.S. Department of Homeland Security to shelve its plans for a Disinformation Governance Board in 2022: Critics raised concerns that the vaguely named entity could threaten free speech and promote inappropriate censorship, despite its much more innocuous mission.[12]

Although some past efforts *have been* unethical and others *seemed likely* to be unethical, it is also clear that at least some influence operations under some circumstances are ethically justifiable. While it is true that noble ends do not always justify every means, a significant part of DoD and the U.S. Department of State's remit is to influence foreign actors in ways that align with both ethically sound and politically advantageous outcomes. For example, encouraging and facilitating human rights programs through truthful and appropriate means in countries with deeply troubling human rights records is not only ethically appropriate but ethically urgent. The task, therefore, is to understand what makes some influence programs, such as these, ethically permissible and others impermissible.

How is DoD organized to evaluate the ethical appropriateness of potential influence efforts as part of its concept of operations (CONOPS) approval process? Unfortunately, this is another area rife with challenges. According to interviews for this project and years of experience interacting with practitioners, it is our understanding that there are no formal, *explicit* processes for considering the ethics of specific influence operations, and the personnel responsible for planning and executing influence operations do not receive training specific to the ethics of influence. However, ethics are *implicitly* considered in at least two ways during the influence planning and approval process.[13] First, plans and CONOPS are subject to review by DoD lawyers to ensure that they are legal, and many of the practitioners we spoke with equated actions that are deemed *legal* and *consistent with authorities* to being *ethical*.[14] However, our view is that what is legal is not necessarily ethical; we spoke with one DoD lawyer who characterized proposals for actions that are legal but unethical as "lawful but awful."[15] Second, personnel in approval chains (individuals who review and approve or disapprove proposed plans or CONOPS) for influence operations sometimes raise concerns or objections that have an ethical character. These concerns are not always clearly articulated and might reflect an aversion to risk rather than ethical concerns; they might indicate that a proposed course of action (COA) is "wrong" or "unacceptable" or that the optics or appearance of the COA could lead to is "bad." In such cases, genuine ethical concerns might be obscured because the ethical principle being violated has not been specified. These concerns can lead to process delays because approvers and planners lack the right terms and concepts to discuss and resolve these ethical conundrums. Furthermore, mislabeling concerns as related

---

[10] Philip M. Taylor, *Munitions of the Mind: A History of Propaganda from the Ancient World to the Present Day*, 3rd ed., Manchester University Press, 2003, p. 267.

[11] Mazzetti, 2004.

[12] Shannon Bond, "She Joined DHS to Fight Disinformation. She Says She Was Halted by . . . Disinformation," NPR, May 21, 2022; U.S. Department of Homeland Security, "Following HSAC Recommendation, DHS Terminates Disinformation Governance Board," press release, August 24, 2022.

[13] Retired U.S. military officer with experience in influence, interview with the authors, August 17, 2022.

[14] DoD civilian employee with oversight responsibilities in influence, interview with the authors, August 10, 2022.

[15] Retired U.S. military judge advocate with experience in the legal review of influence, interview with the authors, October 3, 2022.

to appearances can conceal real ethical dilemmas, and DoD has obligations to all uniformed and civilian personnel (and all Americans) to structure military operations (including influence operations) in a manner that is ethically sound.

Unlike the ethics of influence, the ethics of force is a prominent topic in military training and education, and it is effectively integrated into the practice of U.S. military personnel. All officers are schooled in the law of armed conflict (LOAC) and understand the related ethical principles of military necessity, humanity, honor, distinction, and proportionality. All service members should understand the importance of knowing and following the rules of engagement (ROE), which govern their use of force at each stage of an operation. However, while troops have the tools to determine whether the application of lethal force is ethical and lawful in any situation they might face, they are not similarly prepared to make ethical determinations regarding influence. For example, is deceiving an enemy combatant more problematic than killing them? It is likely that most U.S. service members have not had to consider or discuss this question. If they had, they might rightly conclude that an enemy liable to physical harm (i.e., legitimately subject to lethal force under the ROE) can also be permissibly deceived (a lesser harm). However, the processes for approving military deceptions are far more cumbersome than the processes for release of weapons under the ROE. What about deceiving civilians? The ethics related to physical violence in warfare condemn the intentional targeting of civilians; in fact, they require that service members accept heightened personal risk to protect civilians. However, it *is* legal and ethical to launch attacks that might cause incidental harm to civilians or collateral damage to civilian structures or vehicles if the attack is directed at a legitimate military target and adheres to other governing principles, such as military necessity and proportionality. Given the prohibition against intentionally targeting civilians but the permissibility of risking civilian harm under certain circumstances, does deceiving (or influencing) civilians count as intentionally targeting them for harm? Under what circumstances is that harm acceptable? Where does the acceptable influence of civilians cross over into impermissible harm? DoD processes and training leave troops better prepared to address and resolve dilemmas related to civilian physical harm than to contend with similar dilemmas related to influence.

Methods and Approach

Paul and Marcellino's decades of combined experience studying DoD influence efforts revealed the relevance of the questions posed in this study and the likely existence of a gap in DoD processes.[16] To confirm

---

[16] See, for example, Paul, 2008; Isaac R. Porche III, Christopher Paul, Michael York, Chad C. Serena, Jerry M. Sollinger, Elliot Axelband, Endy M. Daehner, and Bruce Held, *Redefining Information Warfare Boundaries for an Army in a Wireless World*, RAND Corporation, MG-1113-A, 2013; Christopher Paul, Jessica Yeats, Colin P. Clarke, Miriam Matthews, and Lauren Skrabala, *Assessing and Evaluating Department of Defense Efforts to Inform, Influence, and Persuade: Handbook for Practitioners*, RAND Corporation, RR-809/2-OSD, 2015; Christopher Paul and William Marcellino, *Dominating Duffer's Domain: Lessons for the U.S. Marine Corps Information Operations Practitioner*, RAND Corporation, RR-1166-1-OSD, 2017; William Marcellino, Meagan L. Smith, Christopher Paul, and Lauren Skrabala, *Monitoring Social Media: Lessons for Future Department of Defense Social Media Analysis in Support of Information Operations*, RAND Corporation, RR-1742-OSD, 2017; Christopher Paul Colin P. Clarke, Michael Schwille, Jakub P. Hlavka, Michael A. Brown, Steven Davenport, Isaac R. Porche III, and Joel Harding, *Lessons from Others for Future U.S. Army Operations in and Through the Information Environment*, RAND Corporation, RR-1925/1-A, 2018; Christopher Paul, Colin P. Clarke, Bonnie L. Triezenberg, David Manheim, and Bradley Wilson, *Improving C2 and Situational Awareness for Operations in and Through the Information Environment*, RAND Corporation, RR-2489-OSD, 2018; Ben Connable Michael J. McNerney, William Marcellino, Aaron B. Frank, Henry Hargrove, Marek N. Posard, Rebecca Zimmerman, Natasha Lander, Jasen J. Castillo, and James Sladden, *Will to Fight: Analyzing, Modeling, and Simulating the Will to Fight of Military Units*, RAND Corporation, RR-2341-A, 2018; Christopher Paul, Yuna Huh Wong, and Elizabeth M. Bartels, *Opportunities for Including the Information Environment in U.S. Marine Corps Wargames*, RAND Corporation, RR-2997-USMC, 2020; Michael Schwille, Anthony Atler, Jonathan Welch, Christopher Paul, and Richard C. Baffa, *Intelligence Support for Operations in the Information Environment: Divid-*

this intuition and further specify the problem, we interviewed 19 defense influence subject-matter experts (SMEs) with backgrounds in influence operations or experience with the approval processes for influence operations. More than half of these individuals were already known to us because of prior research on information- and influence-related topics; the remaining interviewees were referred to us by the initial collection of experts.

With a refined understanding of the problem and confirmation of the gap from the SME interviews, we partnered with philosophers familiar with military ethics—Michael Skerker, Jeremy Davis, and Bradley J. Strawser—to identify appropriate principles for ethical influence and develop a framework for their application in the DoD context. Through a series of regular virtual meetings, we shared our knowledge and expertise regarding relevant defense practices and learned about the views and positions of ethics philosophers. These meetings progressed to workshops in which we identified examples and test cases, made individual and collective judgements about those test cases, and then discussed the grounds for our determinations. In this process, we always sought to be as clear as possible about the reasoning underpinning our determinations.

From repeated consideration of a variety of test cases (which covered several possible conditions and types of operations and emphasized "tough" or potentially borderline cases), patterns emerged. We distilled those patterns into principles grounded in ethics philosophy, then we developed a framework and workflow that would allow military personnel to apply those principles.

Finally, we used a scenario-based approach to test the utility of the principle-based assessment workflow. These scenarios, which are fictional but plausible, afforded us the opportunity to test and revise the workflow so that it made both ethical and operational sense. This report presents the results of that process.

Steps Toward a Solution: A Framework for Ethical Influence

As noted previously, DoD influence operations face several challenges related to influence: public concerns regarding the appropriateness of any form of influence, a lack of explicit consideration of ethics in the influence-planning process, and decoupling the ethics of force from the ethics of influence in military operations. How, then, should DoD plan and review proposed influence operations to ensure that they are conducted in accordance with sound ethical principles and, if appropriate, are justified and explained in a way that overcomes these ethical challenges? These questions are the focus of the remainder of this report.

We recommend the adoption of a framework for considering ethics as a part of every proposed influence effort. The report includes

- clear ethical principles that should govern the planning and conduct of influence operations
- clear procedures for assessing ethics and the ethical risk associated with a proposed influence operation
- guidelines for creating a justification statement for a proposed influence operation based on a preliminary ethical determination so that reviewers and approvers are presented with a consistent, coherent, and nonarbitrary ethical evaluation with which they can engage and agree or disagree.

ing Roles and Responsibilities Between Intelligence and Information Professionals, RAND Corporation, RR-3161-EUCOM, 2020; William Marcellino, Christopher Paul, Elizabeth L. Petrun Sayers, Michael Schwille, Ryan Bauer, Jason R. Vick, and Walter F. Landgraf III, Developing, Disseminating, and Assessing Command Narrative: Anchoring Command Efforts on a Coherent Story, RAND Corporation, RR-A353-1, 2021; and Christopher Paul, Michael Schwille, Michael Vasseur, Elizabeth M. Bartels, and Ryan Bauer, The Role of Information in U.S. Concepts for Strategic Competition, RAND Corporation, RR-A1256-1, 2022.

The principles are that influence operations should be necessary, effective, and proportionate. These principles are refined into five criteria: Military influence efforts should (1) seek legitimate military outcomes, (2) be necessary to attain those outcomes, (3) employ means that are not harmful (or harm only those liable to harm), (4) have high likelihoods of success, and (5) not generate second-order effects beyond what is intended. Where any of these criteria are not fully satisfied, the harm implicit in violating that criterion is significantly outweighed by the benefit gained (the principle of proportionality). The framework structuring the application of these principles is intended to unfold in three phases: an initial screening, followed by a full ethical risk assessment (if necessary) and the preparation of a justification statement.

Going forward, joint and service doctrine relevant to influence operations should include consideration of ethical concerns. Planners and practitioners should adhere to the ethical principles presented here and should follow a structured and logical process to make ethical determinations. A process like this one should be formally added to the review and approval procedures so that planners and reviewers can benefit from additional process discipline and clarity.

Note that the framework described in this report recognizes that different individuals might reach different conclusions regarding the permissibility of certain actions. That is expected and perfectly reasonable. This framework and approach promote consensus on process: All planners, practitioners, and reviewers should agree that (1) it is appropriate and even morally necessary to make an ethical assessment of a proposed influence operation and (2) unless some very strict criteria are unambiguously satisfied, it is necessary and appropriate to write a justification statement related to the ethics of the proposed effort. This consensus on process is further supported by a common vocabulary that describes what is and is not acceptable, makes points of disagreement clear, and supports movement toward resolution and consensus for individual proposals.

Outline of the Remainder of the Report

This report follows a *spiral learning model*, which introduces concepts, reinforces them through repetition, then deepens comprehension through additional detail and illustration. While the key concepts have been presented here, they are elaborated and developed in the following chapters. In Chapter 2, we further detail current thinking and practice related to ethics in influence operations; consider input gathered from interviews with practitioners, reviewers, and approvers; and provide a review of the relevant philosophical literature. In Chapter 3, we build on the discussion in Chapter 2 and lay out clear principles for the ethical conduct of influence operations, followed by a framework and process for the ethical evaluation of proposed influence operations. This framework contains three parts: an initial screening that determines whether a proposed effort raises any ethical concerns, followed by a full ethical risk assessment for any but the most unambiguously virtuous of proposed efforts, and then the preparation of a justification statement. To illustrate how the framework might be applied, in Chapter 3 we conclude with several practical scenarios based on influence activities that might occur during an armed resistance to occupation.

CHAPTER 2

Current State of Thinking and Practice Regarding the Ethics of Influence

The first step in our research effort was to get context on the ethics of influence: What is the current practice and understanding of ethics of influence across DoD, and how might contemporary ethical scholarship inform military thinking on the ethics of influence? The rest of this chapter lays out top-level findings from a series of SME interviews we conducted, along with a summary of key insights from ethics scholarship. These findings inform our proposed ethical principles for influence and their application in military operations.

Observations from Practitioners and Defense Officials

To better appreciate the current practices and understanding of the ethics of influence, we conducted semi-structured interviews with 19 influence operations SMEs. These SME interviews included both current and former practitioners, as well as those who had experience at higher echelons in approval chains for influence operations, from across DoD. We found that, in general, the ethics of influence are not explicit in planning, training, decisionmaking, or risk assessment. This is not to say that ethics are not important; on the contrary, all our SME interviews affirmed that the ethics of influence were. However, *ethics specific to influence are not clear and explicit in DoD but are implicit in a variety of adjacent areas.*

An example of this ambiguity is the participants equating ethics with legality, including the LOAC, U.S. law (specifically, the Authorization for Use of Military Force of 2001),[1] legal review by higher echelons, theater ROE, and command authorities at the combatant command level. Participants also pointed to nonmilitary frameworks (e.g., religion, Boy Scouts, personal morals) to guide ethical decisionmaking. And while participants identified general ethics training for military members, they could not point to influence-specific ethics training that they had received or were aware of. Finally, when asked about the ethics of influence, participants often shifted to *reputational risk*: While they could not identify explicit ethical frameworks for influence, they generally worried that influence efforts could go awry and do more reputational harm than whatever good could be derived.

Additionally, the ethics of influence do not appear to be explicit in planning. While participants specifically mentioned COA development, COA analysis, and wargaming, they generally presumed that the analysis of the ethics of influence took place at higher echelons of the military. For example, tactical decisionmaking is complicated, and initiative and tempo must be encouraged. Ethics and oversight for tactical-level influence planning might be bundled into ROE or included through review at the operational level. In this case, however, the ethics (including the ethics of influence) would be presumed to be implicit to the operation, whether through legal review or the assumption that the United States prosecutes only just wars.

---

[1] Public Law 107-40, Authorization for Use of Military Force, September 18, 2001.

There is a disconnect between the ethics of influence operations and the ethics for the use of force: Participants pointed out that the U.S. military is more worried about lying to an enemy combatant than killing them and generally saw this as problematic. One officer stated, "Legitimate targets of military violence should have the same ethical standards [as influence]."[2] One participant argued that while the United States was comfortable with the ethics of killing Iranian General Qassem Suleimani, messages urging others to kill him would never have been approved because they might have led to unforeseen violence. This sense that the effects of information cannot be controlled (but somehow the informational effects of kinetic operations can be) is one reason practitioners and approval chains might hesitate to approve influence operations that are not straightforward efforts to inform.

These issues all make the approval process for influence efforts harder. The lack of an explicit ethical framework, concerns about reputational risk, and the lack of formal ethical considerations in planning also constrain approval. While military planners and approval chains have explicit methods for operational risk management—training, risk assessment matrixes, and risk annexes—none of these methods include managing influence risks. Thus, assessing the risks of influence makes it difficult to develop CONOPS and get them approved. This can trigger a reaction in military planners that results from risk aversion or simply an unwillingness to reach a decision (what participants called "slow-rolling" approvals). These difficulties and complications demonstrate the need for explicit frameworks for thinking about what constitutes ethical conduct in influence and how to assess ethical risks in proposed influence efforts.

Insights from Ethics Scholarship

The prior section described the current practices and understanding of the ethics of influence across DoD based on a series of SME interviews. This section reviews existing ethics scholarship and summarizes key concepts in ethics that are relevant to informing the ethics of influence.[3]

The Moral Challenge: Influence Deprives Targets of Their Autonomy

Influence operations are intended to cause their target audiences to choose actions desired by the influencer. These actions might or might not be actions the target would have chosen on their own without outside influence. Influence operations are morally controversial because of their potential effect on a target's autonomy, which—in some understandings—is the very core of personhood, the root of human rights, and the main value to be protected by law in liberal democracies. Again, *virtuous persuasion* is permissible because it respects the listener's autonomy: The persuader provides accurate information and their good faith opinion with the hope that the listener will choose to agree. However, despite the fact that influence operations are potentially ethically problematic, there are nevertheless cases in which such operations are morally permissible.

In the next subsection, we outline the case against influence operations—that is, the main reasons for skepticism regarding the ethicality of influence operations. This is followed by a subsection that discusses the necessary features and circumstances for such operations to be ethically appropriate.

---

[2] Retired U.S. military officer with experience in influence, interview with the authors, August 4, 2022.

[3] This section deals with some basic ethical concepts accepted by the consensus of moral philosophers. For a review of this foundation, the basic philosophical building blocks of rights and autonomy, see Thomas Hobbes, *Leviathan, or, the Matter, Form, and Power of a Common-Wealth Ecclesiastical and Civil*, 1651; John Locke, *Two Treatises of Government*, 1689; Immanuel Kant, *The Metaphysics of Morals,* 1797; and Immanuel Kant, *The Critique of Practical Reason*, 1788. Where there are contested views or views specific to certain authors, we have included additional citations.

The Case Against Influence Operations

Autonomy literally means "self-legislation." Most adult humans have the capacity to be full-fledged moral agents: They are free, can make their own decisions, and lead their own lives. They can therefore be held morally responsible for those decisions precisely because—and to the extent that they are—autonomous. Unlike most other animals, they can decide whether to act on their desires (e.g., they do not automatically eat whatever food falls near them, like dogs do) and choose what they want to pursue while taking into account other people's rights and interests. If a person wants to buy a loaf of bread, they take into account respect for another's property and purchase the loaf. They do not grab it and run out of the store. Furthermore, this is done freely. We would blame a thief who steals the loaf and does not take others' rights into account, but we would not blame someone if they were coerced (for instance, if they were forced at gunpoint to take the bread from the store). Also, we would not blame someone who behaved with a lack of agency (for example, if they were sleepwalking when they stole the bread).

The capacity for autonomy is seen by many as the foundation of human rights, which can be thought of as protecting expressions of autonomy in specific areas of thought, speech, or action. The argument goes that because all people have the innate capacity for autonomy, all humans, no matter where they live, what they believe, or how they are treated by their governments, have the same set of natural human rights. If someone has a right to X, they can freely choose how to use X and demand that others respect their rights and choices regarding X's use. Limiting or overriding an innocent person's right without justification is a rights violation.

Deception Violates Autonomy and the Implied Right to Honest Dealing

One of the central ethical concerns about influence operations is that they constitute a serious violation of autonomy. This is because influence operations can—and often do—ignore or override an individual's ability to make their own choices and deliberate rationally about what to do, what to believe, and who to trust, or they produce the desired outcomes by short-circuiting an individual's reasoning through tactics aimed at stoking fear, resentment, and various other nonrational responses. Furthermore, these operations tend to use individuals as a "mere means" to various desired ends,[4] thereby failing to respect their autonomy as well. The wrongness of immoral actions, such as murder, rape, theft, cheating, deception, torture, and kidnapping, can be summarized by these actions treating someone as a mere means—without any regard to the victim's interests, desires, or humanity and without the victim's ability to consent to the treatment.

A key right that is relevant to influence operations and integral to protecting autonomy is the right to honest dealing. Generally, people have a right to be told the truth regarding matters that affect them. This excludes information that does not affect them to which they generally do not have a right (e.g., their neighbor's deepest religious beliefs or sexual preferences). This extends to the truth about who is telling them something: People have a right to know the true source of a message or communication because deceptive attribution of a message that is otherwise true is still a violation of the right to honest dealing.

Deception is a means of infringing on the right to honest dealing and compromising a person's autonomy.[5] A biological parasite, such as a tapeworm, seeks to use its host body to enrich its own. Similarly, deception seeks to enlist the target's faculties to get them to "freely" perform the actions desired by the agent based on false, misleading, or decontextualized information. In a sense, violent coercion (e.g., a gun to the head) is less insidious than deception because the victim knows they are being forced to do something against their will. The deceived party, by contrast, believes that they are acting in their own best interest when they are really doing what is good for the deceiver.

---

[4] Samuel Kerstein, "Treating Others Merely as Means," *Utilitas*, Vol. 21, No. 2, June 2009.

[5] Michael Skerker, *An Ethics of Interrogation*, University of Chicago Press, 2010, pp. 103–108.

Deception works by altering the target's sense of what is true. The target pursues their normal agenda in service of their own—or their associates'—interests, but the deceiver's false information will not lead them to actually meet their goals.[6]

Emotional Appeals and Autonomy

Emotional manipulation works somewhat differently than deception. In this case, the victim is emotionally affected in such a way that—despite having the relevant facts—they value or weight those facts differently in a way that affects the outcome of their reasoning. Emotional manipulation can range from the stimulation of positive feelings, such as love of family, pride, and respect for great deeds, to harshly antisocial emotions, such as rage, resentment, bigotry, and terror. The more positive emotions might lead one to temporarily alter their priorities. A warmhearted commercial showing friends clinking soda bottles around a campfire might subtly affect a consumer's image of that soda and incline them to buy it instead of a rival soda when shopping. The ad does not change the information they have about soda, nor is it likely to turn someone who does not drink soda into a soda drinker. By contrast, a political ad courting racist fears of immigrants might short-circuit a voter's reasoning, blotting out other considerations in favor of—or against voting for—a certain candidate. Fear might lead the voter to overlook or ignore statistics about crime among immigrants or unsavory information about the fearmongering candidate.

Not all appeals to emotion are obviously ethically problematic. It might be important in some cases to appeal to one's emotions in contexts in which reason alone is unlikely to highlight important values. For example, a doctor might find that even after stating all the facts, they are better able to get their patient to understand the risks involved in a medical intervention through emotional appeals. For the same reason, emotional manipulation can also be instrumentally useful in securing ethically better outcomes. For example, showing people pictures of the horrors of violence (rather than using more-sanitized language) might aid in their understanding, despite the pictures appealing mostly to their sense of compassion. Furthermore, such narrative forms as literature and film often highlight certain features of complex scenarios that might be difficult to communicate otherwise.

Of course, these means can also be used in ways that are ethically questionable, as discussed above. Emotional manipulation is problematic insofar as—and to the extent that—it employs means to deliberately undermine or bypass an individual's autonomous choice. Appealing to one's emotion for reasons that are compatible with their autonomous choice, such as drawing on sentiments that they would rationally identify as important and valuable (e.g., "Would you want your mother to know you've done this?"), is much less problematic than stoking emotions that prey on psychological weaknesses that they would, after rational reflection, decline to endorse.

Coercion and Threats

Influence operations might also try to change people's behavior through threats, whether implied or actual. Standard examples of threats (e.g., the mugger's "Your money or your life!") consist of a coerced choice between an undesirable action that the victim would not have performed voluntarily otherwise (and perhaps one for which other means of persuasion would not have been as successful) and a much more undesirable alternative outcome in which serious harm (such as violence) will follow. Threats differ from other related mechanisms of encouraging action (such as incentives and ultimatums) because they generally impose potential unjust or unreasonable harms to which the victim is not liable so that the threatener can get what

---

[6] Immanuel Kant, "Of Ethical Duties Towards Others, and Especially Truthfulness," in Peter Heath, ed. and trans., and J. B. Schneewind, ed., *Lectures on Ethics*, Cambridge University Press, 1997; Eliot Michaelson and Andreas Stokke, eds., *Lying: Language, Knowledge, Ethics, and Politics*, Oxford University Press, 2018; Jorg Meibauder, ed., *The Oxford Handbook of Lying*, Oxford University Press, 2018; Sissela Bok, *Lying: Moral Choice in Public and Private Life*, 2nd ed., Vintage, 1999; Bernard Williams, *Truth and Truthfulness: An Essay in Genealogy*, Princeton University Press, 2002.

they want. Threats are problematic because they manifest a profound disrespect of the victim's autonomy, typically by using them as a means to produce actions and outcomes that they would not choose otherwise.

Influence Can Threaten Trust and Respect

Deception and emotional manipulation are also concerning because they undermine trust between people. Nearly every kind of social interaction depends on some level of trust. A will not work with B, date B, vote for B, loan money to B, or befriend B if A cannot trust what B is saying. It is important to see that whether someone is trusted, or whether others view someone as trustworthy, is essentially an entirely subjective evaluation made by the other party. One can act in a way that meets all relevant criteria for being viewed as trustworthy by others (for example, honest dealing, being truthful and transparent, and avoiding conflicts of interest), but this does not guarantee being viewed as trustworthy. Therefore, if someone hopes to be trusted, it is essential for them to identify and signal the features of their character through their approach to engaging with others that are most important for the party whose trust is desired. Ironically, however, this might lead a person to be viewed as less trustworthy because their approach might be viewed as less authentic. Gaining trust, particularly when hostility is a natural default response, is difficult. But even after trust is gained, trust often remains fragile. Losing trust is easily done: Even a seemingly benign misstep or miscommunication can cause relationships to fray, often beyond repair. Regaining trust is even more difficult than gaining trust initially.

For these reasons, deception is particularly corrosive in authority relationships, such as the relationship between people and their government or the relationship between citizens and an occupying military force. The social contract between a democratic government and citizens in which the government has the power to tax, regulate, fine, imprison, and conscript depends on the idea that citizens are really ruling over themselves by electing leaders. This contract is voided if citizens do not have accurate information about what their leaders are doing and do not trust that their votes are being counted.[7]

The Case in Favor of Influence Operations

Influence efforts that deprive targets of some part of their autonomy are morally suspect. Under what, if any, circumstances can influence efforts be justified?

Virtuous Persuasion

First, we note that influence efforts that *do not* deprive the target of their autonomy are not ethically problematic. Influence by reason and argumentation that relies on the unbiased presentation of information believed to be true by the influencer and that does not seek to bypass rational or emotional logic through some form of manipulation is morally sound.[8] Such efforts are arguably reason-enhancing and autonomy-enhancing, providing the audience with more complete information from which to arrive at their own independent and autonomous conclusion. This conclusion is consistent with the preferences of the influencer only to the extent that both the influencer and the influenced make similarly reasoned calculations or because their interests align. This is called *rational persuasion* or *reasoned argumentation* in the literature we reviewed, but we prefer *virtuous persuasion* to emphasize that these are persuasive efforts that do not infringe on the

---

[7]   Christopher Kutz, "Secret Law and the Value of Publicity," *Ratio Juris*, Vol. 22, No. 2, June 2009; Dennis F. Thompson, "Democratic Secrecy," *Political Science Quarterly*, Vol. 114, No. 2, Summer 2009; David Luban, "The Publicity Principle," in Robert E. Goodin, ed., *The Theory of Institutional Design*, Cambridge University Press, 1996.

[8]   Jennifer S. Blumenthal-Barby, "Between Reason and Coercion: Ethically Permissible Influence in Health Care and Health Policy Contexts," *Kennedy Institute of Ethics Journal*, Vol. 22, No. 4, 2012, p. 351.

target's autonomy but include appeals beyond just factual reasoning, such as appeals to emotional reasoning (i.e., not just *logos* but also *pathos*).[9]

Justifying Influence of Armed Combatants

Influence efforts aside from virtuous persuasion might still be permissible. Deception, emotional manipulation, and threats are morally concerning and should always lead operators and planners—at a minimum—to pause prior to undertaking an influence effort relying on these techniques. Deception, manipulation, and threats are not always forbidden but require justification.

One way these actions can become more easily justified is if targets of such efforts are liable to harm (i.e., if they have forfeited their rights). Forfeiture is often raised in the context of permissible violence: Violating or attempting to violate another's rights to life can lead a person to temporarily forfeit their right to life or to not be harmed (e.g., when a militant attempts to set off a bomb in a marketplace). Proportionate and necessary levels of violence can be justified to defend oneself or others against such an unjust attacker.

Legitimate targets of violence are also legitimate targets of deception, manipulation, and threats because these are lesser forms of coercion aimed at interrupting their actions.[10] Terrorists, drug dealers, or pirates have no right to engage in their rights-violating actions and so cannot reasonably complain if a foreign military uses deception, trickery, threats, or emotional manipulation to interfere with their plotting or operations.[11] In some scholars' views, service personnel embarked on an unjust war (see below) have no right to engage in violence and so also forfeit rights against deception, trickery, and the like.

Militaries often engage in ruses, threatening bluffs, incitement, encouragements of defection, and surrender. Participants usually know this when they deploy. Soldiers cannot reasonably complain that the enemy, whom they are trying to trick and kill, tricked them first.[12] Both sides are legitimately seeking an advantage over the other in the interest of the security of their state.[13]

Justifying Influence of Noncombatants

As we have described, operations involving deception or the incitement of strong antisocial emotions can cause harm to armed combatants. Predictably, such operations often harm noncombatants as well. In general, harm to armed combatants is easier to justify than harm to civilians for the reasons discussed above: Specifically, combatants waging an unjust war forfeit their rights against such harms. By contrast, civilians have done nothing that justifies harm; civilians, then, seem off-limits. Both contemporary just war theory and the LOAC hold that targeting civilians for violent harms is categorically prohibited: It is always wrong to intentionally kill or maim civilians.

---

[9] Our use of *virtuous* denotes persuasion that is morally appropriate and is not intended to have any religious overtones, implications, or connections.

[10] See Skerker, 2010, pp. 14–20; and Arthur Isak Applbaum, *Ethics for Adversaries: The Morality of Roles in Public and Professional Life*, Princeton University Press, 1999, Chapter 7.

[11] Skerker, 2010, p. 164.

[12] Skerker, 2010, pp. 152–155.

[13] One exception to this is *perfidy*: that is, when one side pretends to bargain or act in good faith with the intention of getting the opposing party to let their guard down so that the former can make offensive maneuvers more easily. One example is pretending to surrender so that the enemy draws closer and then attacking them; another is disguising military vehicles as the Red Cross to gain access to vulnerable locations. Perfidy violates customary international humanitarian law because tolerating such acts would significantly erode the practices that allow for surrender and various other humanitarian actions within war. For a further discussion of perfidy, see Arthur Ripstein with Oona A. Hathaway, Christopher Kutz, and Jeff McMahan, *Rules for Wrongdoers: Law, Morality, War*, ed. by Saira Mohamed, Oxford University Press, 2021, pp. 42–52.

While this principle seems true in cases of serious bodily harm or death, is it true for the sorts of harms that might result from influence operations? Although they are less severe in intensity, the harms that influence operations might cause are worth serious moral consideration. Harm to one's autonomy is generally less significant and clearly less severe than the harms from which civilians are typically morally exempt, such as loss of life or serious bodily injury. They are also less severe in duration: The harms caused by influence operations tend to cease or diminish over time after the operation has concluded. Furthermore, the harms are often—though not always—for the benefit of those they harm.

Furthermore, unlike violent bodily harms (most of which are contrary to the goals of the war and make achievement of the mission less likely), influence operations can require targeting civilian groups. Indeed, in many cases, to avoid targeting these groups would make the success of such operations impossible. It is therefore important to ask whether and when it can be justified to intentionally target civilian groups for influence operations.

No one of these reasons on its own justifies intentionally targeting civilians for influence operations. However, together they show that such harms are not categorically forbidden, as is the case with serious bodily harms. They also suggest a set of guidelines for justifying such harms to civilians.

Just War and Ethical Conduct in War

These guidelines stem from the foundational features of just war thinking. Philosophers and theorists have divided questions about the ethics of war into two main categories: *Jus ad bellum* (JAB) governs the ethics of resorting to or entering a war, and *jus in bello* (JIB) governs ethical conduct within a war. Many of the philosophical disputes regarding the ethics of war center on how best to understand the relationship between these two categories and the various criteria that constitute them.

JAB concerns the ethics of resorting to or entering a conflict. According to the standard interpretation of JAB, each of several conditions must be met for entry into war to be ethically permitted. Although there is much philosophical disagreement about the precise set of criteria that must be met, there is a general consensus about the following three conditions—each of which is relevant for our discussion of influence operations.[14] These conditions are as follows:

1. *Just cause.* The state must have one or more of a restricted set of just causes (e.g., a significant and widespread rights violation, such as an unjust attack on a state's people or the state's sovereignty).
2. *Proportionality.* The harms inflicted by the war must not be out of proportion to the goods it will achieve.
3. *Last resort or necessity.* The war must be the option of last resort; all available less harmful alternatives must be reasonably attempted before war is a permissible option.

In addition to the criteria that govern the resort to war, there are distinct but related principles of JIB, which govern conduct within war. These criteria are as follows:

---

[14] For discussions of the foundations of just war theory, see Thomas Nagel, "War and Massacre," *Philosophy & Public Affairs*, Vol. 1, No. 2, Winter 1972; Michael Walzer, *Just and Unjust Wars: A Moral Argument with Historical Illustrations*, Basic Books, 1977; Barrie Paskins and Michael Dockrill, *The Ethics of War*, University of Minnesota Press, 1979; Richard Norman, *Ethics, Killing, and War*, Cambridge University Press, 1995; Brian Orend, *War and International Justice: A Kantian Perspective*, Wilfrid Laurier University Press, 2000; and Brian Orend, *Michael Walzer on War and Justice*, McGill-Queen's University Press, 2001.

1. *Discrimination.* As previously noted, it is a centerpiece of just war thinking that civilians are not legitimate targets, and actions in war must distinguish between legitimate and illegitimate targets and avoid harm to the latter category.

2. *Proportionality.* Whereas the JAB version of proportionality looks at the overall harms-to-benefits ratio across the entire conflict, the JIB version of proportionality looks at a specific act within a war— that is, a specific mission, engagement, or act of violence—and asks whether the harms this action will inflict are in proper proportion to the goods it will achieve.

3. *Last resort or necessity.* Whereas the JAB version of last resort looks at the resort to war itself, the JIB version dictates that any resort to violence within a war (e.g., every act of using violent force, such as shooting and killing) is only permissible if all available less harmful alternatives have been reasonably attempted.

The foregoing principles offer a helpful set of guidelines for how to justify influence operations as well. (Because discrimination has already been addressed in this section, we will set it aside for now.)

Just Cause

First, in the absence of a just cause, harmful influence operations are unjust. That is, if A attacks B unjustly, A's use of influence operations to further its unjust aims is also unjust. This is true regardless of how benign the harms the influence operation will bring about might seem. This point also relates to JAB's proportionality condition. Many hold that without a just cause, there are no meaningful benefits that a war can provide. That side of the equation is therefore zero, and thus the harms will always outweigh the benefits. In other words, proportionality cannot be satisfied in the absence of a just cause.

Moreover, operators should only be acting in their national interest. It is obviously an abuse of power to deceive a foreign intelligence officer (for example, leading them to believe that their teenager is abusing drugs) if it is merely done by the operator for their own amusement.[15]

Even with this fairly loose standard of national security interests, we can still rule out actions that are not designed to defend the inhabitants of one's state or those of allies from violent attack, crime (e.g., piracy or narco-trafficking), or other kinds of coercion (e.g., blockades or cyberattacks). National security interests are narrower than national interests. Actions furthering national interests might include actions that advance a state's economic interests, are meant to terrorize the civilian population of another country, are designed to stoke ethnic or religious violence in another state, or are meant to interfere with another state's democratic processes. These are not the proper domain of military influence operations.

Proportionality

Assuming there is a just cause for a military operation, we can then ask about the proportionality of a given influence operation. The proportionality calculation for a given influence operation is complex because it involves not only immediate harms and benefits but long-term effects as well.

In many cases, deceptive or manipulative influence operations aimed at national security targets that have forfeited or waived their rights should still be ruled out if the second- or third-order effects are severe. In the kinetic realm, the second-order effects of an airstrike on an ammo dump might include the secondary explosions of the enemy's munitions cooking off. Even though the operators might be confident that they can accurately hit the center of the ammo dump and avoid nearby civilian structures, they might not be able to model the trajectories of all the secondary explosions. In this case, third-order effects might include the

---

[15] Applbaum, 1999, Chapter 4; John M. Parrish, *Paradoxes of Political Ethics: From Dirty Hands to the Invisible Hand*, Cambridge University Press, 2007.

effects on families of civilian casualties killed by the second-order effects or the economic effects that follow the infrastructural damage caused by exploding an ammo dump.

In the realm of influence operations, second-order effects might include a loss of confidence in the media if operators place false stories in local newspapers or online. As noted above, this loss of trust can be corrosive to civil society and hinder the ability to foster cooperative relationships for securing moral goods in the future. Trust in social media could degrade if members of a population suspected that some of the posts in their feeds came from foreign military actors. Once again, even the perception of deception can degrade trust among adversaries.

Even the provision of accurate information can have unplanned negative effects. Could comic books, cartoons, or other media designed to discourage recruitment by insurgents lead young people to speak out against insurgents and subsequently lead to harsh crackdowns that the information's originators are in no position to stop? The same question could be asked regarding information (true or false) designed to encourage anti-regime sentiments in authoritarian countries. It seems the height of irresponsibility for operators to whip up anti-regime sentiments by creating the illusion of a groundswell of opposition to the regime. What could these operators do when a handful of influenced persons take to the streets in the face of riot police and realize too late that most people still support the regime? In some cases, spreading accurate but unflattering information about a regime or an insurgent figure could spur a backlash against that figure's ethnic or religious group. In other cases, there is a risk of contamination beyond the target group. Information is global now. Citizens of the operators' own state might believe the disinformation targeted at a foreign group.

In short, the knock-on effects of influence operations need to be carefully considered. The local environment, including political, ethnic, and religious sensitivities, needs to be well understood before an operation is executed. Knock-on effects that include unjust violence against individuals not liable to harm will almost always rule out an influence operation, unless that operation was credibly designed to stop unjust violence on a larger scale. Forestalling imminent violence could justify a more-distant possibility of a loss of trust in various media, but influencing a target population does not usually justify the loss of trust in various kinds of media.

Again, the issue of forfeiture arises with knock-on effects. Because militant group members have forfeited rights not to be deceived, it is not problematic if an influence operation is projected to sow distrust among them as a second-order effect. However, these knock-on effects are problematic if they affect ordinary citizens of a target country. Their rights cannot be discounted simply because of their nationality; they must be weighed equally with the operators' own citizens. A simple test is a variation of the Golden Rule: Would citizens of my country object if they were targeted—or if they were the subject of knock-on effects—by foreign military influence operators?

In addition to weighing the first-, second-, and higher-order effects, the proportionality calculation invites us to consider whether certain harms count more than others. That is, should operators consider harms to their side to have more weight than those to others? This issue is complex, and there is substantial philosophical disagreement about how to answer it. Unlike the lay public, most just war theorists seem to think that whether a harmed party is a conational or compatriot is not particularly morally significant; all lives have equal value, and only factors concerning what an individual has done—that is, whether they have forfeited their rights—matter. The mere fact that someone is or is not my compatriot should not affect what is permissible to do to them. However, other scholars argue that compatriot partiality can make a moral difference, though only in relatively modest ways.[16] Nevertheless, one area of agreement is that whether someone is a compatriot does not override all moral constraints and considerations: The fact that someone shares

---

[16] Jeremy Davis, "Scope Restrictions, National Partiality, and War," *Journal of Ethics & Social Philosophy*, Vol. 20, No. 2, 2021.

a nationality with me might make me prefer their life over that of a non-compatriot, but there are limits to this preference.

What is important about this line of argument is that the fact that those who are harmed are citizens of another country might make these harms somewhat easier to justify than if they were our compatriots, but—even in the more-permissive position—this fact does not make such harms morally insignificant or irrelevant. At the same time, harms suffered by our compatriots count for more: Operators ought to exercise greater caution regarding harms that might befall their own compatriots and should give these harms greater weight in determining whether an operation is permissible.

Last Resort or Necessity

In addition to questions about proportionality, there is the related question of necessity. As it is typically construed, the necessity (or last resort) condition holds that war must be the option of last resort; all less harmful options must be tried before considering violence. This condition also has a JIB variant: Any violent action within a war must be the option of last resort. Because influence operations are not violent in the traditional sense, they might be viewed as satisfying the last resort condition by using a nonlethal option before lethal options are considered. Nevertheless, influence operations are subject to the last resort condition as well. That is, given that influence operations are potentially harmful, less harmful alternatives that achieve the same ends must be considered. More generally, one must ask whether a particular influence operation—and the harms it might engender—is necessary at all.

Conclusion

Deception, emotional manipulation, and threats are morally concerning. At a minimum, operators and planners should pause prior to undertaking an influence effort that relies on these techniques. Deception, manipulation, and threats are not always forbidden but require justification. First, deceptive influence operations should serve a just cause. Second, they must be necessary to accomplish a goal that cannot be achieved with transparent communication. Third, the reasonably anticipated benefit of the operation must significantly outweigh the negative second- and third-order effects on populations who are not liable to harm.

CHAPTER 3

Ethical Principles for Influence and Their Application

Building on Chapter 2, in this chapter, we propose guidance and a structure for considering the ethics of influence and describe clearly applicable principles and processes for doing so using existing ethical scholarship. The chapter begins by distilling principles to guide ethical practices for influence and then lays out a framework composed of three stages: (1) an initial screening, (2) a full ethical risk assessment (if necessary), and (3) the preparation of a justification statement (if, in fact, the planner concludes that the proposed influence effort is justified). This is followed by several illustrative scenarios in which we apply the framework to realistic examples of influence operations.

Principles That Should Guide the Ethical Planning and Conduct of Influence Operations

Several ethical principles should guide the development and approval of defense influence operations. If possible, defense influence operations should not cause harm. If they do cause harm, the harm caused should be constrained to only those who are liable to harm (such as enemy combatants); if that is not possible, the harm to those who are not liable should be minimized. Because defense influence operations can cause harm, we argue that such efforts should be *necessary, effective,* and *proportionate* (building on the principles of just war theory described in Chapter 2).[1] By *necessary,* we mean something consistent with the LOAC concept of military necessity: Necessary actions are not excessive, they must contribute to the attainment of the goal, and no alternative actions are better able to achieve the goal.[2] In other words, if a harmful act is necessary, it should cause no more harm than is required to achieve a legitimate goal. The requirement to be *effective* compels the planner to ensure that the action is as likely as possible to deliver progress toward the goal; actions that are likely to cause harm but are unlikely to produce the desired effect violate this principle. Multiple actions might be roughly equally effective. Considered together, effectiveness and necessity should lead planners to choose the least harmful approach that is still effective. The principle of *proportionality* considers the ethical balance of harm to be inflicted and the advantage to be gained: The harm likely to be caused must not be excessive in relation to the advantage or benefit likely to be gained as the outcome. Note that proportionality does not allow these things to be equal: Causing harm that is equal to the benefit gained is not proportionate. Instead, the likely positive benefit must significantly outweigh the likely harm.

To support the evaluation of these highest-level principles (necessity, effectiveness, and proportionality) in practice, we propose a framework in which users consider five criteria and then weigh them in a proportionality calculation. Military influence efforts should seek legitimate military outcomes, be necessary to

---

[1] Cécile Fabre, *Spying Through a Glass Darkly: The Ethics of Espionage and Counter-Intelligence,* Oxford University Press, 2022, p. 3.

[2] For further discussion, see Nobuo Hayashi, "Requirements of Military Necessity in International Humanitarian Law and International Criminal Law," *Boston University International Law Journal,* Vol. 28, No. 1, 2010.

attain those outcomes, employ means that are not harmful (or harm only those liable to harm), have high likelihoods of success, and not generate second-order effects beyond what is intended. When a proposed effort falls short of fully achieving one or more of these five criteria, likely harm should fall predominantly on those who are liable to harm and should be proportionate. In other words, the positive benefits of military influence efforts should substantially outweigh the harm they cause. We discuss these criteria in greater detail and the proposed approach to proportionality calculus in the following sections.

A Framework for Applying Ethical Principles for Influence in Operational Planning

Drawing on a framework that is familiar across DoD for assessing risk in operational planning, we present a practical framework for practitioners and approvers to determine the ethical permissibility of a planned or proposed influence operation.[3] Planners and commanders must balance potential risks and harms against operational ends for all operations—including influence operations. Because it is a familiar conceptual anchor, we use risk management assessment to build a workflow and templates that are similar to the tools used in exercises or at operational-level commands (i.e., non-doctrinal templates). The framework and templates are provisional and might need iteration and improvement, but planners should aspire to apply the framework's *principles* even if the described process does not fit cleanly within the workflow in their command.

The framework is not a calculator or black box that spits out yes-or-no ethical determinations. Rather, it is a structured approach that allows practitioners to reason through a planned operation, reach individual determinations, then share those determinations in a way that allows others (for example, a reviewer or approver) to follow their logic and agree or disagree. It is important to note that two users of the framework might reach two different conclusions about whether a proposed operation is ethical. However, the framework should help users identify what ethical principle is at stake and what risk assessment regarding that principle is the source of disagreement. Ultimately, this would allow users to engage in *productive* disagreement, whether that is reaching a consensus, finding a way to adjust the proposed activity so that their assessments agree, or including both assessments (and the reasons for the disagreement) in the approval package so that the commander can make an informed decision.

How the Framework Is Intended to Work: A Three-Stage Application

The proposed framework for the ethical evaluation of influence efforts is intended to unfold in three phases: an initial screening, followed (if necessary) by a full ethical risk assessment and the preparation of a justification statement. Figure 3.1 shows the workflow for the framework. When a planner is ready to consider the ethics of their proposed influence operation, they should initiate Stage 1, ask themselves the three initial screening questions, and make a supporting heuristic assessment (described in greater detail in the following subsections). If this initial screening raises absolutely no concerns (that is, all screening questions are answered with a resounding "no," and all heuristics are clearly in the affirmative), then the proposed influence operation is unambiguously ethical and can be implemented without ethical concerns. If, however, any of the Stage 1 questions elicited an affirmative, uncertain, or qualified response—or if any of the heuristic assessments reflect any level of ethical uncertainty—then the planner should proceed to Stage 2. Progressing to Stage 2 does not necessarily mean that a proposed action is unethical. Rather, it simply means that

---

[3] Department of the Army Pamphlet 385-30, *Risk Management*, Headquarters, Department of the Army, December 2, 2014; Marine Corps Order 3500.27C, *Risk Management*, Department of the Navy, November 26, 2014.

FIGURE 3.1

Workflow of the Application of the Framework for Ethical Influence

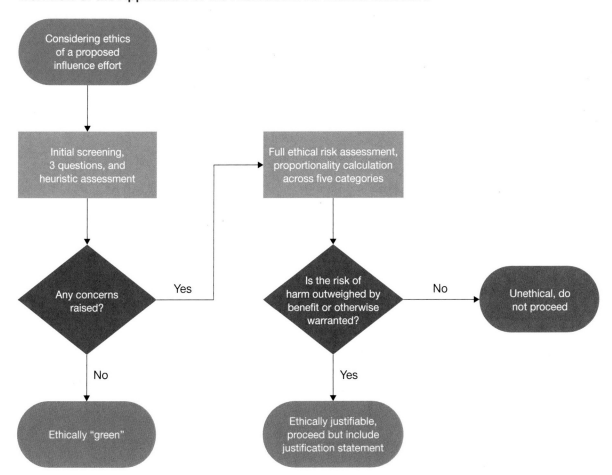

the proposed effort requires more careful and more thorough consideration. Whether a proposed action is determined to be permissible or impermissible, the ethical considerations and details of the proportionality calculation should be captured in a justification statement.

Stage 1: Initial Screening

As discussed in Chapter 2, manipulative or deceptive influence is a violation of the target's autonomy and thus a form of harm. Because there are many circumstances that could allow for DoD to undertake harmful influence operations, a full ethical risk assessment is recommended if harm is an expected result of an operation, moving its evaluation immediately to Stage 2. However, some influence efforts fall into the realm of virtuous persuasion and are not in any way harmful. All public affairs activities fit into this category, for example. Influence through virtuous persuasion is not harmful and thus raises no ethical concerns. To see whether a proposed effort belongs in this category, our framework offers three screening questions and three heuristic tests. The three screening questions are as follows:

1. Is any part of the proposed effort deceptive?
2. Does any part of the effort manipulate audiences (emotionally or socially)?
3. Given the intended and plausible outcomes, will (or might) anyone be harmed?

Is any part of the proposed effort deceptive? Inclusion of one or more rhetorical fallacies might bring an affirmative answer to this question, as should any content that is biased, spun, or misleading. Even content that is true but involves lying by omission should trigger a positive answer here. This also applies to attribution. If the source of the content provided as part of the proposed influence effort is not clear, then the answer to this question should be "yes" or "maybe" and trigger progress to Stage 2.

Does any part of the effort manipulate audiences (emotionally or socially)? Positive and prosocial emotions (hope, loyalty, inspiration, kinship, etc.) and negative emotions (fear, anger, guilt, etc.) can be useful in thoughtful, reasoned consideration of an argument leading to behavior change and thus can be ethical. However, if the influence effort relies on invoking strong negative emotions (e.g., existential fear, hatred), it might compromise the audiences' reason and merit a Stage 2 evaluation as potentially unethical. Similarly, manipulating an audience through social means (e.g., invoking long-standing ethnic hatreds) might compromise audiences' ability to make a thoughtful decision about being persuaded.

Given the intended and plausible outcomes, will (or might) anyone be harmed? Even if the influence effort is honest and avoids emotional appeals or manipulation, what might happen because of the effort? This includes both intended consequences (the desired outcome) and unintended consequences (e.g., collateral damage). For example, if the influence effort intends to induce peaceful protest in a target audience by truthfully detailing the underhanded dealings of an adversary nation's government in a regional radio broadcast, then the possibility of violent protest that such a broadcast inspires and the adversary government's potential response with violent crackdowns means that someone might be harmed because of this action. Therefore, such an action demands further consideration in Stage 2.

In the case that all three preliminary questions are answered in the negative, practitioners should consider three complementary heuristic tests (if any of the initial questions were answered "yes" or were ambiguous, users can proceed to Stage 2). These heuristic tests are intended to expose any residual doubt or uncertainty that deserves further exploration in a full ethical risk assessment (Stage 2).

Heuristic A: *the reciprocity test.* Ask yourself, "Would you be okay with someone using this approach on your grandmother (or other elderly relative)?" If you object or are not sure, proceed to Stage 2. Sometimes, the actions taken on distant others that do not seem harmful or deceptive would trigger a different reaction when we consider their application a little closer to home.

Heuristic B: *the* New York Times *test.* "If a journalist found out about what you are proposing, would it be newsworthy, and would DoD be okay with the resulting article?" If what you are proposing is something that is routine and perfectly clean from an ethical perspective, then a journalist would be unlikely to be interested, or, if they did write a story, no one in your chain of command would care. If, however, an article or exposé about this proposed operation is something that might cause consternation or blowback, then it deserves full consideration in Stage 2.

Heuristic C: *the supervision test.* "What would your boss or commander say about this if they were looking over your shoulder?" Would your boss have provided the same answers you did to the three initial screening questions, or might they assess the operation differently? If you are not 100 percent confident that your boss would be 100 percent okay with this proposal, then it merits a full ethical risk assessment in Stage 2.

Figure 3.2 provides a worksheet for the initial screening process. Again, if any of the first three questions elicit a "yes," proceed to the full ethical risk assessment. Likewise, if any of the three heuristic tests elicit a "no," proceed to the full ethical risk assessment.

FIGURE 3.2

Example of an Initial Screening Worksheet

Ethical Risk Assessment: Initial Screening Worksheet

Proposed activity: _____

1   **Is any part of the effort deceptive?**

Does the effort include any of the following:

Misleading content (misleading framing of an issue)	Y/N?
Imposter content (impersonating genuine sources)	Y/N?
Fabricated content	Y/N?
False or no attribution	Y/N?
False connection (headlines or visuals don't support other content)	Y/N?
False context (genuine content with false contextual information)	Y/N?
Manipulated content (genuine imagery or content manipulated to deceive)	Y/N?

2   **Does any part of the effort manipulate audiences socially or emotionally?**

What emotions could the effort create? _____

Are any of the emotions created strong, negative emotions?	Y/N?
Is there social manipulation in the effort, such as invoking group membership or ethnic tensions?	Y/N?

3   **Given the intended and plausible outcomes, might anyone be harmed?**

Who is intended to receive the effort? _____

Who might receive content unintentionally? _____

Could anything bad or harmful happen to anyone because of this effort?	Y/N?
Does the activity violate anyone's autonomy, their freedom to make free choices?	Y/N?

If any of the above questions were answered YES, proceed to the Full Ethical Risk Assessment

If any of the below questions are answered NO, proceed to the Full Ethical Risk Assessment

A	**Reciprocity test: "Would you be okay with someone using this approach on your grandmother (or other elderly relative)?"**	**Y/N?**
B	***New York Times* test: "If a journalist found out about what you are proposing, would it be newsworthy and would DoD be okay with the resulting article?"**	**Y/N?**
C	**Supervision test: "If your boss or commander were looking over your shoulder, would they agree with your assessment?"**	**Y/N?**

Stage 2: Full Ethical Risk Assessment

In Stage 2 of the application of the framework for ethical influence operations, the user makes a full ethical risk assessment—considering the proposed effort's relative virtues and including the harm or benefit it might cause and whether any of the harm is warranted (or whether the targets are liable to such harm) across the five evaluative criteria—then compiles these findings into a single proportionality calculation to make a final assessment. The five criteria and their application—rightness of the intended outcome, necessity, rightness of means, likelihood of success, and higher-order consequences—are described below. Each criterion follows from one or more of the ethical principles (necessity, effectiveness, and proportionality). Rightness of the intended outcome supports the principle of necessity; for something to be necessary, it must be necessary for a legitimate purpose. The necessity criterion supports the principle of necessity. Rightness of means supports the principle of proportionality; if the means employed in an influence activity are harmful, that harm must be appropriately proportionate to the expected benefit. Likelihood of success supports the principle of effectiveness. For a proposed influence effort to be deemed effective, it must have reasonable prospects for success. Likelihood of effectiveness also supports the principle of proportionality; if the odds of an effort's success are low, then the possible gains must be very great or the amount of likely harm must be negligible or very low to be proportionate. Finally, the criterion of higher-order consequences supports the principle of proportionality because it requires the consideration of downstream effects that might cause harm that would also need to be outweighed by the expected benefit.

Legitimate Intentions

The first consideration is whether the intended outcome of the proposed influence operation is legitimate. This is an assessment of whether a cause is just and—if it is—whether the effects or outcomes the proposed influence effort seeks to achieve will contribute to the broader ends sought. If your country is at war and it is a just war,[4] then outcomes that lead toward your side prevailing (provided that achieving them respects other ethical requirements) are legitimate. Likewise, gaining certain kinds of military advantages during wartime contributes to legitimate military objectives. During a just war, inflicting physical harm on enemy combatants is legitimate and arguably good because it limits the enemy's ability to resist and contributes to the resolution of the conflict.

Clearly legitimate military objectives during a just war are the low-hanging fruit for the rightness of an intended outcome. The legitimacy of military influence objectives will follow the legitimacy of the use of force *during wartime*, and all officers in the joint force should be familiar with the rules governing the use of force. The legitimacy of objectives is less obvious when operations are below the threshold of warfare and at other points on the competition continuum.[5]

If the justification that "this is a legitimate military objective as part of a just war" is unavailable, then an ethical consideration needs to identify some other grounds for claiming a just cause for generating influence. Many such justifications are possible. Though not necessarily ethically equivalent to a legitimate military outcome during a just war, the authority under which forces operate might include a justification for the use of force that might specify legitimate military objectives that could also be legitimately pursued through influence. Some possible justifications include the following:

- Self-defense and the defense of others are valid justifications. While they are more commonly used in contexts involving physical harm, they could also be used to justify influence efforts to prevent any form of harm.

---

4 See the discussion of JAB and JIB in Chapter 2.

5 JP 3-0, *Joint Operations*, Joint Chiefs of Staff, August 11, 2011.

- If an adversary or competitor undertakes operations that violate human rights or other rights, then efforts to stop them from doing so and to restore rights are justified.
- If an adversary or competitor threatens to begin operations that violate human rights, other rights, or the sovereignty of another nation, then efforts to prevent them from doing so can be justified. However, it is important to note that efforts at deterrence rest on firmer moral ground than preemption.
- If a group of individuals with a *moral obligation* to complete or prevent some action does not carry out or stop it, then operations that aim to get that group to fulfill its duty (or to meet the obligation on the group's behalf) can be justified.
- Counteracting other actors' wrongful influence efforts by restoring the wrongfully influenced to their original attitudes and behaviors can be justified.

In applying this criterion, remember that there are two levels that need to be considered for the rightness of the intended outcome. At the first level, are the overall ends just? At the second level, if so, do the outcomes sought by the proposed influence effort contribute to those ends? Note that even in a clearly justifiable operational context (such as a just war), objectives that have nothing to do with the overall mission do not satisfy this criterion. For example, selfish efforts to defraud local civilians and enrich oneself or an effort to get revenge on an ex-spouse who works for a partner-nation government are not acceptable, even if pursued during a justified war.

Note that if this criterion is not satisfied, then the evaluation is over: As it stands, the proposed effort is not ethical. An effort that does not contribute to a legitimate objective cannot be justified and should be abandoned or substantially redesigned.

Consider the following example: Seeking the surrender of enemy forces is a legitimate military objective during a just war. However, there are other ethical criteria to consider because of other aspects of a proposed operation. If the plan is to promote surrender by relentlessly shelling an enemy position for a period, then broadcasting a surrender appeal by loudspeaker, it is unlikely to run afoul of any other ethical criteria. If the goal is to encourage an enemy formation to surrender by influencing the families of the formation's soldiers to pass on a surrender appeal and share how well captives are treated, the surrender of enemy troops remains a legitimate objective, but the approach will need to be carefully evaluated based on other criteria. If the goal is to encourage an enemy formation's troops to surrender by capturing their families and threatening them, that approach is almost certainly going to be found to be unethical because it violates other criteria, even though the objective (surrender promotion) remains legitimate.

Necessity

This is the only one of the three higher-level principles (necessity, effectiveness, and proportionality) that translates directly as one of the constituent criteria of the evaluative framework. Necessity relates to the legitimacy of an objective when evaluated according to the previous criterion: Is the proposed action necessary to bring about the intended outcome? Are there other ways to achieve the same effects or otherwise meet the objective? If there are other possible ways, then is there an argument that the proposed influence effort is in some way better (or at least no worse) than the other ways? That is, is it less costly in terms of time or resources or less dependent on a scarce resource (such as a low-density, high-demand asset), or is it less harmful or does it pose less risk of harm to other people besides the targets of the effort? Even if a proposed action meets the standard of being better than the alternatives, does it cause more harm than is required? Necessary actions must achieve or contribute to a legitimate goal, be the only or the best way to achieve that goal, and cause no more harm than what is required to be effective.

To evaluate necessity, a practitioner should consider other ways the joint force might achieve the same objectives or outcomes expected from the proposed influence effort and the relative costs and merits of those actions. In fact, the COA development and analysis step of the Joint Operation Planning Process, service-level Marine Corps Planning Process or Military Decision Making Process, could serve as a template.[6]

Note that this criterion must also be satisfied to proceed. The proposed COA must be at least as good as all other COAs when considered against all COA evaluation criteria. If there is a better COA available, then the only ethical thing to do would be to use that other approach.

Ethical Means

This criterion asks whether the means of influence being proposed are ethical. *Means* includes the rhetorical approach and the kind of appeal being made (whether that involves misleading or manipulating the target and thus depriving them of autonomy), as well as the behavior or attitude being influenced and the intended change (whether the intended behavior constitutes harm to the target or to others). Means also includes who is targeted for influence and the mode and media of the influence effort (which affects how precise targeting is likely to be), who undertakes the means (such as members of the armed forces, government civilians, contractors, or others), and whether they are identified as the source of the means (known as *attribution*).

As discussed in Chapter 2, virtuous persuasion is always an ethical means. Also note that while telling the truth has virtue, it can also be used to mislead (partial truth) or cause harm (malinformation). For example, sharing the truth with a population about their political leader's corruption could lead to protests, violence, or other forms of harm. This does not mean that revealing such information is not permissible, but if someone might be harmed as a result of the disclosure, then thoughtful consideration is required. Physical deceptions (such as feints, demonstrations, and ruses) that do not use explicit lies but allow observers to draw their own (mistaken) conclusions are also generally permissible. However, such influence efforts still need to consider whether they cause harm to someone who is not liable to harm (such as a civilian family who evacuates and flees from a feint and finds themselves relocating into the actual route of advance). Furthermore, the permissibility of physical deceptions stops unequivocally at perfidy, which is never acceptable.

The means of a proposed influence effort could be ethically problematic in several ways. First, as noted previously, deceptive or manipulative means are violations of individuals' presumed autonomy, so the application of such techniques is harmful. This extends to efforts that involve false pretenses, such as content that is portrayed as independent journalism but is actually paid content. Second, if the change in behavior sought is harmful, either to the influenced or to others, then the effort is harmful. Remember that *harm* as an ethical consideration is broad; beyond physical harm, it includes harm to welfare, autonomy, dignity, trust, and freedom. An example of harm to the influenced might be a "come out and fight" appeal that challenges the target's masculinity to make them easier to bring under fire; an example of harm to others is a "rise up" appeal during a resistance that leads to violence and harm among both the incited and the occupiers. Third, if the influence effort is at all harmful, the target of the influence activities could be problematic: Are the targets all liable to harm, or have they forfeited rights to not be harmed in this way? Fourth, the level of precision of the means can be problematic. Wide area broadcasts or leaflet drops might reach many in a target audience but might also reach a broad secondary audience. If an influence effort is at all harmful, then more *discriminating* means are preferred to reduce the total number of individuals subject to harm and to avoid harming those who are not liable to harm. Harm to an unintended audience is particularly concerning. Imagine, for example, if the aforementioned "come out and fight" appeal was heard and heeded by youths, or others outside the intended target audience, many of whom are incited to engage the joint force (and be engaged in return). Fifth, the actor tasked with carrying out the operation can raise ethical concerns. If the activities of an influence effort are conducted by a legitimate representative of a government or military in pursuit of legitimate military or policy objectives, then the user of the means is not problematic. However, if the means

---

[6]   JP 5-0, *Joint Planning*, Joint Chiefs of Staff, December 1, 2020.

are harmful, and private citizens carry out influence actions outside an official capacity, or such actions are undertaken by contractors, mercenaries, or cutouts, then the effort becomes much harder to justify because of their lack of authority or accountability. Sixth, if the influencer does not fully and accurately identify themselves (or their government affiliation), that can also be a harmful form of deception. Displaying content that is intentionally anonymous (with an anonymous byline or a clear pen name) is honest to the extent that a receiver knows that the sender has chosen not to reveal their identity and can imagine a variety of possible motives for that choice. Falsely attributing such content is at least minimally harmful because it deprives the audience of the ability to accurately reason about the motives and credibility of the sender. Claiming a false attribution could potentially be even more harmful if the assumed identity or persona is a real person or a real organization: The misattributed content might cause the person or organization to which it is ascribed to suffer harm, whether minimal harm to their reputation or physical or economic harm in the form of reprisal. Falsely attributing content to a real individual or organization can quickly escalate to possible harm.

If the means are harmful in any way, the influence effort is presumptively wrong unless the harm can be justified. Broadly, there are two types of reasonable justifications for harmful means: (1) The targets are liable to harm or have forfeited related rights (either because they are legitimate military targets or because they have a moral duty or obligation and are thus liable to efforts to persuade them to meet that obligation), and (2) the harm caused is minimal and overwhelmingly offset by the likely benefits of the effort (which should be weighed in the overall proportionality calculations at the end of this process).

Likelihood of Success

The likelihood of success ties directly to the principle of effectiveness, which is a requirement for ethical influence operations. This criterion considers the extent to which the proposed effort is likely to deliver the effects sought. This is evaluated as a likelihood rather than a certainty because there is always some uncertainty and contingency in military operations, including influence operations. Uncertainty about success can come from many sources, and a planner making a Stage 2 ethical framework evaluation for the likelihood of success should consider them carefully. Uncertainty about success can come from uncertainty about execution. An influence activity that is difficult to do (because there is difficulty with access, a challenge related to translation, or a contingency from third-party delivery) can have a reduced likelihood of success. Alternatively, uncertainty can stem from the genuinely variable effectiveness of the effort: Perhaps only certain members of the target audience are likely to receive or notice the influence effort, or maybe the nature of human dynamics leads different people to respond to a message in different ways (and we lack perfect understanding of exactly why), or perhaps the persuasiveness or virality of the message varies.

Planners can use past experiences to be more confident in their estimates of the likelihood of success of their efforts. This confidence can come from previous similar efforts (for example, they might consider the success of another force's similar influence effort in a different region), pilot tests, or prototype implementation. A history of success or a successful test does not guarantee the same positive result, but it provides a firmer foundation for estimating the likelihood of success.

In a proportionality calculation, the likelihood of success should be thought of as a contributing factor to the positive benefit an influence effort is expected to produce. If the benefit sought is large but the likelihood of realizing that benefit is low, the overall expected benefit is relatively modest. For example, consider an effort to influence the behavior of a single important enemy commander; a change in behavior could result in an important advantage for the joint force, but getting that behavior change would require the influence operation to go perfectly and is predicated on several assumptions about the targeted commander being correct. Such an operation is likely only ethically justified if the likely harm is negligible and the resources required for an attempt are low.

Note that a proposed influence effort that has a probability of success that is not substantially different from zero (that is, it has basically no chance of success) fails this criterion because it fails to adhere to the principle of effectiveness and should be redesigned so that it has a nonzero chance of success. A proposal that has a small but non-negligible chance of success can continue through the evaluation process because a small possible benefit might still meet all other criteria and prevail in a proportionality calculation in which the likely benefit is low but the likely harm and risk are even lower.

Second-Order Consequences

Ethical considerations related to second-order consequences consider harm that might be generated by the knock-on effects of either the actions undertaken as part of influence operations or the intended outcome and effects of these efforts. These can include intended and anticipated effects as well as unintended and unanticipated effects. Unanticipated effects are one of the many forms of risk that are relevant when planning influence operations and are why the proportionality calculation of a proposed influence effort's ethical evaluation feels like a sort of risk analysis: In many respects, it is.

There are factors in generating unintended second-order consequences from influence efforts that stem from *unintended audiences*: someone other than the target receiving the message or signal that was intended to influence. This can happen when an effort lacks *precision* and is broadcast, delivered, or displayed using media available to people outside the target audience. It can also happen when an effort *escapes* its initial medium or broadcast and is shared or rebroadcast by receivers. This could happen in different ways depending on the medium initially used: If the influence efforts involve email or SMS messages sent to specific individuals, those individuals could forward the message to others via the same media. If the original medium is a handbill or leaflet, an observer could photograph it using their phone and post it to social media, giving something originally targeted locally a potential global reach.

Reaching audiences outside the intended audience with an influence effort can be ethically problematic if harm is caused. When considering rightness of means, a planner evaluates whether the intended target of the effort is liable to harm or has forfeited rights that are violated by the influence effort, but what about unintended recipients? If people outside the target audience also receive and are affected by the influence, then they need to be considered as part of the proportionality calculation. While enemy combatants and adversary government personnel might be liable to harm, incidental recipients typically are not. If many individuals who are not liable to harm are even slightly hurt by an influence effort targeting others, then the level of benefit expected from the effort would have to highly outweigh that harm in a proportionality calculation.

Note that there is an analogy for higher-order consequences from influence available in the physical violence of warfare. The same logic that makes physical collateral *damage* ethical and acceptable under certain circumstances during armed conflict might also be used to justify some leakage of an influence effort onto individuals who are not themselves liable to manipulation or deception.

Another form of second-order consequence can stem from unintended outcomes. For example, if a deception intended to remain undiscovered is revealed or if an unattributed message is subsequently identified as coming from DoD, this exposure can cause a loss of trust or produce other problems. Producing the intended behavior can also provoke other behaviors, either by the intended audience or by other actors. For example, efforts to incite nonviolent protests might succeed, but those protests might provoke government repression. Furthermore, efforts to incite a protest might succeed too well and cause a riot or insurrection. Similar to considering the likelihood of success, considering unintended effects is like a risk analysis. What could happen, how bad would the consequences be, and how likely is that bad outcome? To complete a full ethical risk assessment, a planner should identify and consider a variety of possible unexpected but anticipatable outcomes and weigh their likelihood and consequences.

Proportionality

The application of the criterion of proportionality is the point at which all the outcomes of the other criteria are weighed together. For an operation to be justifiable, no criterion can be completely failed and, on balance, the likely benefit derived from the proposed action must substantially outweigh the harm that might be caused to those who are not liable to harm.

Proportionality might be thought of as an equation in which the likely benefit must be substantially greater than the harm when a proposed action does not completely satisfy all criteria. This might look something like the following:

likely benefit from necessary action for a legitimate outcome	*>>*	*likely harm from means, harm to unintended audiences, harm from higher-order consequences*	*–*	*harm to those who are liable to harm.*

This calculation requires some personal judgement and art, but the goal is to make a thoughtful determination that the user believes is correct and to be able to share the grounds for that determination with others as part of a justification statement. A justification statement will allow approvers to agree or disagree with a planner's determination with reference to specific criteria; the disagreement can then be discussed and resolved, or the plan can be adjusted to mitigate issues associated with those specific criteria.

Figure 3.3 presents a sample Full Ethical Risk Assessment Worksheet, which is intended to help a planner review the considerations related to each criterion, keep track of their assessments, and capture the relevant elements in a justification statement. We include an example of a worksheet that is filled out later in this chapter.

Note that the worksheet includes a "stop" symbol (⊘) next to some of the criteria. These criteria need to be satisfied for a proposed action to have any chance of being justifiable. Influence actions that do not contribute to a legitimate objective, that are not necessary, or that have basically no chance of success or expected return cannot be justified. In any of these cases, the proposed action needs to be reconsidered, and planners must go back to the drawing board.

After considering the relevant criteria and completing a worksheet, a user should have enough information to make a preliminary determination: Is the benefit likely to be gained appropriately proportionate to the harm likely to be caused? If so, then the effort is (at least preliminarily) ethically justifiable. The user should proceed to Stage 3: the preparation of a justification statement.

Stage 3: Justification Statement

The completion of the Stage 2 ethical evaluation helps the user reach their preliminary determination regarding the ethical justifiability of a proposed operation. If it is not justifiable, then the proposal needs to be reworked or rejected. If it is ethically justifiable, the proposal still needs to be accompanied by a justification statement. A justification statement will allow reviewers or approvers to follow the chain of logic that uses the relevant ethical criteria and leads to a determination that the likely benefit significantly outweighs the likely harm of an influence operation and then either agree with that determination or dispute it. Should a reviewer not agree with the initial determination, a clear justification statement should help them identify where and why they disagree (for which of the criteria), which is more likely to lead to a productive and specific conversation about that disagreement (rather than vague concerns or nebulous disapproval).

FIGURE 3.3
Example of a Full Ethical Risk Assessment Worksheet

Full Ethical Risk Assessment Worksheet

Proposed activity: _____

1 **Legitimate intentions**

Desired effect (also list intermediate military, operational, or campaign objective that the effect supports):

_____

Is the supported objective a legitimate military objective? Y/N If no, what else justifies this objective?

_____
_____

Will the desired effect contribute to the objective? Y/N If not legitimate or no contribution to effect, 🚫

2 **Necessity**

What other actions or approaches could produce the same effect? _____

_____

Why is the proposed approach better than alternatives? _____

_____ If it is not better, then 🚫

3 **Ethical means**

What parts of the effort are untrue, misleading, misattributed, or emotionally manipulative? _____

_____

Describe any threat to autonomy or dignity _____

What media will be used, and how precise are they?_____

Who will conduct each aspect of the effort? _____

What harm, intended or otherwise, will be caused? How much?_____

If there is harm from this effort, how might it be justified? _____

4 **Likelihood of success**

Likelihood that the proposed activity could be successfully executed/performed: ___%

Likelihood that a sufficient number of target audience receive/perceive the effort: ___%

Likelihood that receiving targets change behavior as desired: ____%

On what evidence or experience are these estimates based? _____

Overall prospects for success (multiply %s as decimals) ___ x ___ x ___ = ___% If success % is near 0%, 🚫

5 **Second-order consequences**

List possible second-order/downstream consequences you have identified (including accidental audiences and unintended effects): _____

Assessment of risk of harm from higher-order consequences:

_____

_____

Summarize all likely harm from **3** & **5** : _____

Overall determination: _____

| *likely benefit from necessary action for a legitimate outcome* | >> | *likely harm from means, harm to unintended audiences, harm from higher-order consequences* | − | *harm to those who are liable to harm* |

To prepare a justification statement, follow any guidance or templates used in the relevant process. If the user lacks clear guidance or a template, write a summary of the criteria evaluation worked through in Stage 2. Mention each of the criteria and how the proposed effort satisfies those criteria. Where the proposed effort falls short of a criterion, include the reason that a form of harm or a criteria violation is justified, and summarize the proportionality calculation in words. Consider including a cleaned-up version of the completed Full Ethical Risk Assessment Worksheet as an addendum to the justification statement.

Illustrative Examples of Assessing Ethical Risk in Influence Operations

This report lays out a rigorous, repeatable way to assess the ethics of proposed influence operations. The following examples are meant to bring this assessment framework to life for practitioners through illustrative scenarios, which are somewhat analogous to tactical decision games. The discussion begins with a brief description of a fictional context: A U.S. partner, Pinelandia, has an overbearing and aggressive adversary or competitor neighbor, Krasnovia.

The Ethics of Influence: Resistance Scenarios

The authoritarian regime in Krasnovia seeks to exercise hegemonic control over bordering states—which act as a kind of security buffer zone against Western nations. Krasnovia uses a wide variety of state power instruments in pursuit of this goal but has increasingly turned to direct military action against neighboring states. This has included support of armed separatist groups in border areas within neighboring countries and the use of Krasnovian special forces to create a pretext for the illegal annexation of a border area in Pinelandia, a nonaligned nation that shares a border north of Krasnovia.

Six months ago in late spring, two Krasnovian Army (KA) groups consisting of mechanized, armored, artillery, and airborne brigades, also supplemented by KA special operations units, crossed the Ptushkya river, invading Pinelandia. The Pinelandia Defense Force (PDF) was overwhelmingly outnumbered but showed tremendous will to fight, trading space for time over two months of intense conventional warfare. Currently, the KA has occupied multiple border regions populated with a mix of ethnic Pinelandians and ethnic Krasnovians and has installed local puppet administrators loyal to the Krasnovian regime. As the PDF prepares a counteroffensive, Pinelandia engages in a whole-of-society effort to resist the occupation.[7] The U.S. and Western allies are assisting Pinelandia, including materiel support; training; communications; cyber capabilities; and intelligence, surveillance, and reconnaissance support.

As part of the resistance, various PDF and Pinelandia civilian government units and agencies are actively engaged in influence operations in an effort to bolster Pinelandia's military and civil populace's will to fight, degrade Krasnovian will to fight, and influence global audiences to support their cause. Given the close cooperation between U.S. forces and the PDF, the ethics of Pinelandia's influence operations and any supporting U.S. influence operations are highly salient to U.S. interests and reputation.

Scenario: "We're Watching You"

PDF Special Operations Command (the "Outland Hunters") will coordinate with underground cells in occupied Pinelandia to send a message to potential collaborators: "We're Watching You." Physical graffiti stencils and digital copies of the phrase "We're Watching You" will be distributed and used in occupied territory.

---

[7] For more on whole-of-society resistance to invasion, see Otto Fiala, Kirk Smith, and Anders Löfberg, "Resistance Operating Concept (ROC)," working paper, Joint Special Operations University, May 1, 2020.

Additionally, handbills and posters with the slogan over a picture of a man in a hooded sweatshirt, face obscured, will be distributed. Billboards in border areas will also be deployed.

Discussion: We think this proposed effort requires a full screening because it triggers one of the three screener questions:

1. *Is any part of the proposed effort deceptive?* No, there is no deception involved.
2. *Does any part of the effort manipulate audiences (emotionally or socially)?* This is a fear-based appeal intended to stop those considering collaboration from doing so for fear of future consequences. So, yes.
3. *Given the intended and plausible outcomes, will (or might) anyone be harmed?* Yes, it is quite plausible this operation might inspire reprisal attacks against those who have collaborated or merely been accused of doing so.

Full Ethical Risk Assessment: We think this proposed effort is ethical but clearly involves risks of harm (Figure 3.4). Thus, this risk assessment should lead to a justification statement that acknowledges the risks, indicates why the planner believes the risks are justified, and allows the commander to make a risk-informed decision. Consider each of the five criteria included in the Full Ethical Risk Assessment:

1. *Legitimate intentions.* Pinelandia is engaged in a just war, defending its sovereign territory against an invading and occupying army. Discouraging collaboration and strengthening the will to fight of the entire populace are legitimate wartime aims.
2. *Necessity.* While this is not the only possible means of strengthening the nation's will to resist the invaders, part of the PDF's strategy for resistance involves reducing collaboration with Krasnovia and increasing friction for Krasnovian forces in occupied areas, and doing so requires putting social and personal pressure on potential collaborators in occupied areas.
3. *Ethical means.* As a means, these warning messages are somewhat harmful because they are a fear-based appeal. However, warning potential collaborators that treachery will have negative consequences is a truthful, fair message that is justified by their choice to collaborate. Furthermore, the means are discriminant, aimed at a particular audience (those considering collaborating with the enemy), and justified because only those who have betrayed their obligations to the nation (or are considering doing so) are targets.
4. *Likelihood of success.* The effort is reasonably likely to be successful. Historical efforts to warn and dissuade potential collaborators show some degree of success, and social pressure and clear reinforcement of norms are established social processes. In other words, the content of the campaign is built on a firm understanding of human dynamics.
5. *Second-order consequences.* Some critical possible second-order consequences for this effort are (1) reprisal attacks against collaborators (real or suspected) and (2) false allegations. In the first case, the effort might be interpreted as encouraging vigilante efforts by local populations instead of lawful efforts to charge and try collaborators. In the second case, it is possible that false allegations might be made, which would subject the accused to grave potential harm. *These are serious risks that merit special attention and justification when seeking approval for the effort.*

Scenario: "Operation Phone Home"

This CONOPS has two lines of effort (LOEs). The first LOE targets Krasnovian soldiers in Pinelandia by using both calls and texts to remind them that invaders who commit war crimes face lifetime imprisonment or execution; that soldiers who surrender will be fed, clothed, and treated humanely; that the Krasnovian

FIGURE 3.4

Example Full Ethical Risk Assessment Worksheet for "We're Watching You" Scenario

Full Ethical Risk Assessment Worksheet

Proposed activity: *"we're watching you" billboards, graffiti, handbills, web posts, etc. in occupied areas.*

1 **Legitimate intentions**

Desired effect (also list intermediate military, operational, or campaign objective that the effect supports):
Reduce collaboration with occupiers by making potential collaborators more hesitant, actual collaborators rethink.

Supports LOE 2: Reduce collaboration.

Is the supported objective a legitimate military objective? Y/N If no, what else justifies this objective?
Yes. Krasnovia is an illegitimate occupying power, and activities that increase friction for them are legitimate. Pinelandians have an obligation to resist.

Will the desired effect contribute to the objective? Y/N If not legitimate or no contribution to effect, ⊘

2 **Necessity**

What other actions or approaches could produce the same effect? *Collaborators could be contacted directly, but it would be difficult to identify potential collaborators. Notices could be posted of future intent to prosecute collaborators. Direct action could be taken to harm collaborators.*

Why is the proposed approach better than alternatives? *None of the other options are less ethically risky, but some of them are more difficult or resource intensive.* If it is not better, then ⊘

3 **Ethical means**

What parts of the effort are untrue, misleading, misattributed, or emotionally manipulative? *The effort is only metaphorically true—PDFs aren't actually watching potential collaborators all the time. The appeal is fear/threat based.*

Describe any threat to autonomy or dignity *Fear-based appeal is meant to dissuade collaboration, which people might choose if left alone.*

What media will be used, and how precise are they? *Imprecise blanket messaging. Only members of the target audience need to fear.*

Who will conduct each aspect of the effort? *Some PDF special operations forces, some resistance cells, possible civilian copycat.*

What harm, intended or otherwise, will be caused? How much? *Potential or actual collaborators are subjected to fear, lose some autonomy.*

If there is harm from this effort, how might it be justified? *Targets are liable to mild harm because of failure to meet obligations to not collaborate.*

4 **Likelihood of success**

Likelihood that the proposed activity could be successfully executed/performed: *95* %

Likelihood that a sufficient number of target audience receive/perceive the effort: *95* %

Likelihood that receiving targets change behavior as desired: *5–20* %

On what evidence or experience are these estimates based? *Prior experience with handbills/graffiti, historical anti-collaboration.*

Overall prospects for success (multiply %s as decimals) *.95* x *.95* x *.10* = *9* % If success % is near 0%, ⊘

5 **Second-order consequences**

But that is ~10% per recipient, and a large number of potential collaborators will get this, so success chance is more like 90%.

List possible second-order/downstream consequences you have identified (including accidental audiences and unintended effects): *(1) Reprisals against collaborators, (2) false allegations and reprisals against those falsely accused.*

Assessment of risk of harm from higher-order consequences: *Moderate risk of physical harm to collaborators, minimal risk of harm to innocents. May need a supporting effort to deter mob justice.*

Summarize all likely harm from **3** & **5** : *Likely, warranted minor harm to collaborators; possible unwarranted severe harm to collaborators; possible severe harm to innocents.*

Overall determination: *Permissible if the reduction in collaboration outweighs the risk of unwarranted severe harm.*

| *likely benefit from necessary action for a legitimate outcome* | >> | *likely harm from means, harm to unintended audiences, harm from higher-order consequences* | − | *harm to those who are liable to harm* |

regime and its officers are using them as disposable pawns; and warning them that when they die, their families will have to bear the burden of burying them.

The second LOE targets Krasnovian military families with calls and texts telling them their son is surrounded, hungry, and suffering from cold weather; telling them that their son will likely die and be sent home in a coffin; and asking them to encourage their sons and brothers to surrender and receive humane treatment and repatriation. To further heighten the effect of the effort, the second LOE will include sending graphic pictures of dead Krasnovian soldiers to families identified as having children killed in action in the war.

Discussion: We think this proposed effort fails the initial screener, because while the effort is not deceptive, it emotionally manipulates both soldiers and their families and is likely to cause emotional distress to both audiences. We note that Krasnovian soldiers are liable to harm by virtue of being active combatants in a legitimate military effort. However, we do not find that the families of these soldiers are liable in the same way.

Given this assessment, we think that the effort requires a full screening, and when screened, it will be deemed unethical. First, the means are unethical: Sending pictures of dead bodies to parents to scare them is emotionally manipulative and is likely to cause harm to noncombatants. Second, there are potential second-order consequences: What if family members, in fear for their child, speak out against the war? They might then be subject to arrest, state violence, or vigilante reprisals from Krasnovian nationalists. *In our judgment, the first LOE should be forwarded for approval, while the second LOE should be dropped as clearly unethical.*

Scenario: "High-Value Targets"

PDF regular army and special operations have managed a coup in the past 48 hours: An Outland Hunter sniper team was able to kill Maj. Gen. M---- (Commander, 3rd KA Guards), followed by the capture of Lt. Gen. S---- (Supreme Commander, KA Group North). A quick-approval social and traditional media campaign has been requested to publicize this win with the intent of strengthening Pinelandian and Western will to fight and degrading Krasnovian morale. While both events are real, operators from Special Group 1 have taken steps to heighten the potential impact of the campaign: (1) digitally altering the pictures released of Maj. Gen. M----'s body to be more gory and dramatize his death, and (2) curating capture and post-capture images of Lt. Gen. S---- so that he appears to crawl, plead with raised cuffed hands, and look undignified and unsoldierly to Krasnovian audiences.

Discussion: This proposed effort raises questions related to multiple ethical considerations, including the possible harm of displaying graphic images of death and truthfulness. Further, the proposed effort risks harm to the reputation of Pinelandian forces, although this is not an ethical consideration.

We think that showing death and gore passes three ethical criteria: legitimacy, necessity, and likelihood of success. Vividly showing the cost of war during the war seems fair game for influence meant to end the war. However, digitally altering the pictures does not clearly pass the ethical means criteria because it is untruthful and could have second-order consequences, such as emotional harm to Gen. M----'s family or undue emotional distress to other viewers. (Arguably, they are liable to exposure to actual gore because of their nation's war, but not embellished gore.)

While it is not as untruthful as the effort to digitally alter pictures, we think that the second part of the effort—curating images to tell a particular, partial version of the truth—does not pass the criteria of ethical means. Furthermore, Gen. S---- was a prisoner of war (POW) when some of the pictures were taken, and the intent of this effort is to humiliate a POW, which is unethical and likely in violation of the Geneva Convention. (Note that it would be a different story if Gen. S---- were still a combatant at-large when the humiliating imagery was captured.)

This CONOPS has potential to shift the will to fight in favor of Pinelandia's defense, but it also has significant ethical problems. Digitally manipulating images takes something that is easily justified (disseminating images) and makes it harder to justify, and humiliating a POW is unethical and illegal. Also, if digital alterations are revealed, the risk of blowback is significant in terms of the credibility and reputation of Pinelandian forces. *We would not recommend advancing either element of the CONOPS further in the approval process without some amendment to the proposal. If the gory images were unaltered or if footage of Gen. S---- were confined to pre-capture imagery, the operation would be less objectionable. Whether the original elements of this CONOPS or a revised version is forwarded for approval, it must include a detailed discussion of the ethical risks and potential harm so that the commander and other approvers can make an informed decision.*

Note that this is the sort of scenario that could easily lead different users of the framework to different conclusions. Perhaps one user thinks it is acceptable to show humiliating footage of Gen. S---- because they are not considering the obligations owed to POWs, while another user might remember those obligations and (rightly) have reservations. If both used the framework, they could very easily identify the source of their disagreement and reach either a new agreement or an understanding by which they agree to put the question to a judge advocate general.

Scenario: "Bridge over the River Ptushkya"

The PDF counteroffensive southward in the Newland area of responsibility (AOR) has been successful, pushing back KA forces into a semicircular perimeter defense at the border, centered on the Øresbord bridge. The Øresbord bridge is the main supply route (MSR) between Krasnovia and Pinelandia, with side-by-side highway and railway spans. If the bridge can be destroyed (cutting KA Group North off from resupply), then the PDF should be able to engage and destroy the invaders. This will allow the PDF to then swing eastward and engage KA Group West in a pincer action, which is critical to Pinelandia Supreme Command's (PSC's) national liberation plan.

This CONOPS centers on the physical destruction of the bridge using a high-yield vehicle-borne improvised explosive device, conducted by guerilla cell members in the area. Simultaneous with this strike are two influence efforts. First, cell members will broadcast live dash-cam footage of the detonation from the vehicle and rebroadcast the recordings later, with an agile distribution plan to disseminate to international media outlets. Second, guerilla cell members and Pinelandia Foreign Services representatives have been in contact with anti-regime Krasnovian groups that will be conducting multiple peace demonstrations in major Krasnovian cities that day.

Note that while the Øresbord bridge effort is first and foremost kinetic with the objective of severing the MSR, kinetic effects can have important influence effects as well, and we would stress to readers that *not all "influence efforts" will be purely informational.* The anti-regime peace demonstrators do not know of the bridge attack, but they might face serious state criminal sanctions and potential police violence after the attack footage goes public, which is important to the influence ethics in this case.

Discussion: We feel this CONOPS raises a serious ethical problem in terms of possible second-order effects. While the influence dimensions here likely pass the first four criteria in our full ethical risk assessment, it appears quite likely that civilians *who are working in good faith to promote the end the war* might suffer serious harm in the form of political repression and harm. *We think this effort should NOT go forward for approval unless the planned coordination with anti-regime Krasnovian groups (without their knowledge and consent) is removed or changed to better protect individuals not liable to harm.*

Scenario: "Total Resistance"

The PDF counteroffensive has been successful, and KA Group North is no longer operationally relevant. With the southwest of the country liberated, the PDF are pressuring KA Group East. KA Group East is by far the stronger of the two army groups involved in the invasion and has been reinforced by the Prokofiev Brigade, a military contracting group with strong financial ties to the Krasnovian Regime. This will be a difficult fight.

The Krasnovian regime has shifted its narrative efforts, saying its forces are not retreating but rather accomplishing what was their goal from the start: the rescue of ethnic Krasnovians from planned genocide by PDF "Nazi Kill Squads." PDF and Western intelligence reporting suggests instead that the Krasnovian regime is planning to conduct an ethnic cleansing campaign, forcibly relocating ethnic Pinelandians across the border into Krasnovia, leaving behind an ethnic buffer zone.

In response, PSC is orchestrating a true whole-of-society resistance operation: guerrilla cells, underground cells, resistance auxiliary, and the mobilization of the entire civilian populace to make the region as "sticky" as possible to slow down forced relocation until the PDF can arrive. Messaging directed at the broader civilian populace asks them to resist in every way short of direct violent action. The influence campaign is anchored in the narrative "Hold on as Hard as You Can," with messages such as "Resist. Pull wires and slash tires. Put sugar in gas tanks. Have breakdowns and accidents at intersections. When forced by the occupiers to do *anything*, move slow, get it wrong, drag it out. Resist." Broad resistance in the AOR is critical to Pinelandia's liberation strategy. Successful resistance, particularly in the context of direct action by guerilla units, has a high likelihood of Krasnovian retaliation against the populace.

Discussion: The kind of whole-of-society resistance this scenario portrays raises multiple ethical questions—for example, the broader ethics of enlisting civilians to act as an auxiliary to support guerilla and underground cell operations. For our purpose here, we will focus on the ethics of the "Hold on as Hard as You Can" influence effort. The variety of suggested resistance activities creates substantial risk of potential harms to civilians: Getting caught cutting wires and slashing the tires of military vehicles carries much higher risks than dragging your feet while filling out forms. We will therefore assume that this operation poses great risk of harm: *Is it ethical to ask the civilian populace to resist surreptitiously but with unequivocal acts of resistance?*

Our evaluation of this operation by each criterion is as follows:

1. *Legitimate intentions.* Resisting the invasion of sovereign territory and preventing ethnic cleansing within that sovereign territory are both legitimate miliary strategic goals.

2. *Necessity.* Within the scenario, we think that whole-of-society resistance is urgently necessary; failure to do so increases the risk of successful ethnic cleansing, a great wrong worth taking some risks to stop.

3. *Ethical means.* Although it is a very serious choice, asking the civilian populace to resist and sabotage their own deportation through nonviolent means seems to use ethical means. There is nothing perfidious about this strategy, the means are discriminate and target combatants, the influence effort comes from the legitimate civil authority, and while some of the resistance activities might cause direct harm, they are likely justified given the existential threat to the nation and local populace.

4. *Likelihood of success.* In this scenario, while planners likely cannot accurately predict how much time this effort will buy, it seems likely to have meaningful effects and thus some level of success.

5. *Second-order consequences.* This is where most ethical concerns arise. It is quite possible that civilians who resist and are caught will be harmed and that harm will be severe (for example, if Krasnovian senior leaders decide to treat such civilians as spies and execute them). Furthermore, second-order effects might include harm to nonresisting civilians in the form of mass reprisals. In essence, this is a bold stroke in response to an existential threat and thus poses potentially severe second-order harm risks that are hard to quantify. *Therefore, while we think this effort is ethical, the second-order risks must be laid out in detail and be considered in the approval process.*

Conclusions and a Way Ahead: Adopting and Implementing a Framework for Ethical Influence

In this report, we laid out a theory of ethical influence efforts derived from research on ethics in war. We then synthesized the theory with expert insight, from both influence practitioners and those in influence-approval chains, to generate a set of principles for conducting influence operations ethically. This helped us craft a specific evaluation process with step-by-step guidance to assess the ethical risks of a proposed influence operation in context to guide both influence practitioners and those in approval chains in better risk assessment. While the primary audience for this work is DoD personnel, we recognize that there are adjacent U.S. government and civilian audiences who want to think about what it means to conduct influence operations in an ethical manner that reflects U.S. values.

Going forward, joint and service doctrine relevant to influence operations should explicitly include the consideration of ethical concerns. Planners and practitioners should adhere to the ethical principles presented here and should follow a structured and logical process to make ethical determinations. An assessment like the one we propose should be formally added to the review and approval process to increase process discipline so that planners and reviewers have clear steps to follow. If the workflow and worksheets presented here are adopted, they might need revision and optimization to fit into actual processes. The ethical principles espoused here are time-tested; the worksheets are prototypes.

Adopting such a framework for evaluating the ethical risks of influence operations will produce several important benefits. First, with such a framework in place, DoD influence efforts will be more likely to be ethical and moral. Second, concerns with the ethicality are more likely to be clearly articulated by all participants in the process, including planners, reviewers, approvers, and commanders. Consequently, these concerns are more likely to be resolved (either by reaching a shared understanding or changing the proposed activities to reduce concerns), and more proposed influence operations are likely to be approved with fewer delays and more likelihood of being ethical and moral.

Full-Page Versions of the Initial Screening Worksheet and the Full Ethical Risk Assessment Worksheet

Ethical Risk Assessment: Initial Screening Worksheet

Proposed activity: _____

1   **Is any part of the effort deceptive?**

Does the effort include any of the following:

Misleading content (misleading framing of an issue)	Y/N?
Imposter content (impersonating genuine sources)	Y/N?
Fabricated content	Y/N?
False or no attribution	Y/N?
False connection (headlines or visuals don't support other content)	Y/N?
False context (genuine content with false contextual information)	Y/N?
Manipulated content (genuine imagery or content manipulated to deceive)	Y/N?

2   **Does any part of the effort manipulate audiences socially or emotionally?**

What emotions could the effort create? _____

Are any of the emotions created strong, negative emotions?	Y/N?
Is there social manipulation in the effort, such as invoking group membership or ethnic tensions?	Y/N?

3   **Given the intended and plausible outcomes, might anyone be harmed?**

Who is intended to receive the effort? _____

Who might receive content unintentionally? _____

Could anything bad or harmful happen to anyone because of this effort?	Y/N?
Does the activity violate anyone's autonomy, their freedom to make free choices?	Y/N?

> If any of the above questions were answered YES, proceed to the Full Ethical Risk Assessment

> If any of the below questions are answered NO, proceed to the Full Ethical Risk Assessment

A	**Reciprocity test: "Would you be okay with someone using this approach on your grandmother (or other elderly relative)?"**	**Y/N?**
B	**_New York Times_ test: "If a journalist found out about what you are proposing, would it be newsworthy and would DoD be okay with the resulting article?"**	**Y/N?**
C	**Supervision test: "If your boss or commander were looking over your shoulder, would they agree with your assessment?"**	**Y/N?**

Full Ethical Risk Assessment Worksheet

Proposed activity: _____

1 **Legitimate intentions**

Desired effect (also list intermediate military, operational, or campaign objective that the effect supports):

_____

_____

Is the supported objective a legitimate military objective? Y/N If no, what else justifies this objective?

_____

_____

Will the desired effect contribute to the objective? Y/N If not legitimate or no contribution to effect, ⊘

2 **Necessity**

What other actions or approaches could produce the same effect? _____

_____

Why is the proposed approach better than alternatives? _____

_____ If it is not better, then ⊘

3 **Ethical means**

What parts of the effort are untrue, misleading, misattributed, or emotionally manipulative? _____

_____

Describe any threat to autonomy or dignity _____

What media will be used, and how precise are they? _____

Who will conduct each aspect of the effort? _____

What harm, intended or otherwise, will be caused? How much? _____

If there is harm from this effort, how might it be justified? _____

4 **Likelihood of success**

Likelihood that the proposed activity could be successfully executed/performed: ___%

Likelihood that a sufficient number of target audience receive/perceive the effort: ___%

Likelihood that receiving targets change behavior as desired: ____%

On what evidence or experience are these estimates based? _____

Overall prospects for success (multiply %s as decimals) ___ x ___ x ___ = ___% If success % is near 0%, ⊘

5 **Second-order consequences**

List possible second-order/downstream consequences you have identified (including accidental audiences and unintended effects): _____

Assessment of risk of harm from higher-order consequences: _____

_____

Summarize all likely harm from **3** & **5** : _____

Overall determination: _____

likely benefit from necessary   >>   *likely harm from means, harm to unintended*   −   *harm to those who*
action for a legitimate outcome   *audiences, harm from higher-order consequences*   *are liable to harm*

Abbreviations

AOR	area of responsibility
COA	course of action
CONOPS	concept of operations
DoD	U.S. Department of Defense
JAB	*jus ad bellum*
JIB	*jus in bello*
JP	joint publication
KA	Krasnovian Army (notional)
LOAC	law of armed conflict
LOE	line of effort
MSR	main supply route
OIE	operations in the information environment
PDF	Pinelandia Defense Force (notional)
POW	prisoner of war
PSC	Pinelandia Supreme Command (notional)
ROE	rules of engagement
SME	subject-matter expert

References

Applbaum, Arthur Isak, *Ethics for Adversaries: The Morality of Roles in Public and Professional Life*, Princeton University Press, 1999.

Blumenthal-Barby, Jennifer S., "Between Reason and Coercion: Ethically Permissible Influence in Health Care and Health Policy Contexts," *Kennedy Institute of Ethics Journal*, Vol. 22, No. 4, 2012.

Bok, Sissela, *Lying: Moral Choice in Public and Private Life*, 2nd ed., Vintage, 1999.

Bond, Shannon, "She Joined DHS to Fight Disinformation. She Says She Was Halted by . . . Disinformation," NPR, May 21, 2022. As of November 11, 2022:
https://www.npr.org/2022/05/21/1100438703/dhs-disinformation-board-nina-jankowicz

Connable, Ben, Michael J. McNerney, William Marcellino, Aaron B. Frank, Henry Hargrove, Marek N. Posard, S. Rebecca Zimmerman, Natasha Lander, Jasen J. Castillo, and James Sladden, *Will to Fight: Analyzing, Modeling, and Simulating the Will to Fight of Military Units*, RAND Corporation, RR-2341-A, 2018. As of May 16, 2023:
https://www.rand.org/pubs/research_reports/RR2341.html

Davis, Jeremy, "Scope Restrictions, National Partiality, and War," *Journal of Ethics & Social Philosophy*, Vol. 20, No. 2, 2021.

Department of the Army Pamphlet 385-30, *Risk Management*, Headquarters, Department of the Army, December 2, 2014.

Fabre, Cécile, *Spying Through a Glass Darkly: The Ethics of Espionage and Counter-Intelligence*, Oxford University Press, 2022.

Fiala, Otto, Kirk Smith, and Anders Löfberg, "Resistance Operating Concept (ROC)," working paper, Joint Special Operations University, May 1, 2020.

Frowe, Helen, *Defensive Killing: An Essay on War and Self-Defence*, Oxford University Press, 2014.

Graphika and Stanford Internet Observatory, *Unheard Voice: Evaluating Five Years of Pro-Western Covert Influence Operations*, Stanford Digital Repository, August 24, 2022. As of November 8, 2022:
https://purl.stanford.edu/nj914nx9540

Hayashi, Nobuo, "Requirements of Military Necessity in International Humanitarian Law and International Criminal Law," *Boston University International Law Journal*, Vol. 28, No. 1, 2010.

Hobbes, Thomas, *Leviathan, or, the Matter, Form, and Power of a Common-Wealth Ecclesiastical and Civil*, 1651.

Jenkins, Ryan, Michael Robillard, and Bradley Jay Strawser, eds., *Who Should Die? The Ethics of Killing in War*, Oxford University Press, 2018.

Joint Publication 3-0, *Joint Operations*, Joint Chiefs of Staff, August 11, 2011.

Joint Publication 3-04, *Information in Joint Operations*, Joint Chiefs of Staff, September 14, 2022.

Joint Publication 5-0, *Joint Planning*, Joint Chiefs of Staff, December 1, 2020.

JP—*See* Joint Publication.

Kant, Immanuel, *The Critique of Practical Reason*, 1788.

Kant, Immanuel, *The Metaphysics of Morals*, 1797.

Kant, Immanuel, "Of Ethical Duties Towards Others, and Especially Truthfulness," in Peter Heath, ed. and trans., and J. B. Schneewind, ed., *Lectures on Ethics*, Cambridge University Press, 1997.

Kerstein, Samuel, "Treating Others Merely as Means," *Utilitas*, Vol. 21, No. 2, June 2009.

Kutz, Christopher, "Secret Law and the Value of Publicity," *Ratio Juris*, Vol. 22, No. 2, June 2009.

Locke, John, *Two Treatises of Government*, 1689.

Luban, David, "The Publicity Principle," in Robert E. Goodin, ed., *The Theory of Institutional Design*, Cambridge University Press, 1996.

Marcellino, William, Meagan L. Smith, Christopher Paul, and Lauren Skrabala, *Monitoring Social Media: Lessons for Future Department of Defense Social Media Analysis in Support of Information Operations*, RAND Corporation, RR-1742-OSD, 2017. As of May 10, 2023:
https://www.rand.org/pubs/research_reports/RR1742.html

Marcellino, William, Christopher Paul, Elizabeth L. Petrun Sayers, Michael Schwille, Ryan Bauer, Jason R. Vick, and Walter F. Landgraf III, *Developing, Disseminating, and Assessing Command Narrative: Anchoring Command Efforts on a Coherent Story*, RAND Corporation, RR-A353-1, 2021. As of May 16, 2023:
https://www.rand.org/pubs/research_reports/RRA353-1.html

Marine Corps Order 3500.27C, *Risk Management*, Department of the Navy, November 26, 2014.

Mazzetti, Mark, "PR Meets Psy-Ops in War on Terror," *Los Angeles Times*, December 1, 2004.

McMahan, Jeff, *Killing in War*, Oxford University Press, 2009.

Meibauer, Jörg, ed., *The Oxford Handbook of Lying*, Oxford University Press, 2019.

Michaelson, Eliot, and Andreas Stokke, eds., *Lying: Language, Knowledge, Ethics, and Politics*, Oxford University Press, 2018.

Nagel, Thomas, "War and Massacre," *Philosophy & Public Affairs*, Vol. 1, No. 2, Winter 1972.

Norman, Richard, *Ethics, Killing, and War*, Cambridge University Press, 1995.

Orend, Brian, *War and International Justice: A Kantian Perspective*, Wilfrid Laurier University Press, 2000.

Orend, Brian, *Michael Walzer on War and Justice*, McGill-Queen's University Press, 2001.

Parrish, John M., *Paradoxes of Political Ethics: From Dirty Hands to the Invisible Hand*, Cambridge University Press, 2007.

Paskins, Barrie, and Michael Dockrill, *The Ethics of War*, University of Minnesota Press, 1979.

Paul, Christopher, *Information Operations—Doctrine and Practice: A Reference Handbook*, Praeger Security International, 2008.

Paul, Christopher, "Is It Time to Abandon the Term Information Operations?" *Strategy Bridge*, March 11, 2019.

Paul, Christopher, Colin P. Clarke, Michael Schwille, Jakub P. Hlavka, Michael A. Brown, Steven Davenport, Isaac R. Porche III, and Joel Harding, *Lessons from Others for Future U.S. Army Operations in and Through the Information Environment*, RAND Corporation, RR-1925/1-A, 2018. As of May 16, 2023:
https://www.rand.org/pubs/research_reports/RR1925z1.html

Paul, Christopher, Colin P. Clarke, Bonnie L. Triezenberg, David Manheim, and Bradley Wilson, *Improving C2 and Situational Awareness for Operations in and Through the Information Environment*, RAND Corporation, RR-2489-OSD, 2018. As of May 16, 2023:
https://www.rand.org/pubs/research_reports/RR2489.html

Paul, Christopher, and William Marcellino, *Dominating Duffer's Domain: Lessons for the U.S. Marine Corps Information Operations Practitioner*, RAND Corporation, RR-1166-1-OSD, 2017. As of May 16, 2023:
https://www.rand.org/pubs/research_reports/RR1166-1.html

Paul, Christopher, Michael Schwille, Michael Vasseur, Elizabeth M. Bartels, and Ryan Bauer, *The Role of Information in U.S. Concepts for Strategic Competition*, RAND Corporation, RR-A1256-1, 2022. As of May 10, 2023:
https://www.rand.org/pubs/research_reports/RRA1256-1.html

Paul, Christopher, Yuna Huh Wong, and Elizabeth M. Bartels, *Opportunities for Including the Information Environment in U.S. Marine Corps Wargames*, RAND Corporation, RR-2997-USMC, 2020. As of May 10, 2023:
https://www.rand.org/pubs/research_reports/RR2997.html

Paul, Christopher, Jessica Yeats, Colin P. Clarke, Miriam Matthews, and Lauren Skrabala, *Assessing and Evaluating Department of Defense Efforts to Inform, Influence, and Persuade: Handbook for Practitioners*, RAND Corporation, RR-809/2-OSD, 2015. As of May 16, 2023:
https://www.rand.org/pubs/research_reports/RR809z2.html

Porche, Isaac R., III, Christopher Paul, Michael York, Chad C. Serena, Jerry M. Sollinger, Elliot Axelband, Endy M. Daehner, and Bruce Held, *Redefining Information Warfare Boundaries for an Army in a Wireless World,* RAND Corporation, MG-1113-A, 2013. As of May 10, 2023: https://www.rand.org/pubs/monographs/MG1113.html

Public Law 107-40, Authorization for Use of Military Force, September 18, 2001.

Ripstein, Arthur, with Oona A. Hathaway, Christopher Kutz, and Jeff McMahan, *Rules for Wrongdoers: Law, Morality, War,* ed. by Saira Mohamed, Oxford University Press, 2021.

Rodin, David, *War and Self-Defense,* Oxford University Press, 2002.

Schwille, Michael, Anthony Atler, Jonathan Welch, Christopher Paul, and Richard C. Baffa, *Intelligence Support for Operations in the Information Environment: Dividing Roles and Responsibilities Between Intelligence and Information Professionals,* RAND Corporation, RR-3161-EUCOM, 2020. As of May 16, 2023: https://www.rand.org/pubs/research_reports/RR3161.html

Skerker, Michael, *An Ethics of Interrogation,* University of Chicago Press, 2010.

Taylor, Philip M., *Munitions of the Mind: A History of Propaganda from the Ancient World to the Present Day,* 3rd ed., Manchester University Press, 2003.

Thompson, Dennis F., "Democratic Secrecy," *Political Science Quarterly,* Vol. 114, No. 2, Summer 1999.

Thomson, Scott K., and Christopher E. Paul, "Paradigm Change: Operational Art and the Information Joint Function," *Joint Force Quarterly,* Vol. 89, 2nd Quarter 2018.

Tzu, Sun, *The Art of War,* trans. by Samuel B. Griffith, Oxford University Press, 1963.

U.S. Department of Homeland Security, "Following HSAC Recommendation, DHS Terminates Disinformation Governance Board," press release, August 24, 2022.

Walzer, Michael, *Just and Unjust Wars: A Moral Argument with Historical Illustrations,* Basic Books, 1977.

Williams, Bernard, *Truth and Truthfulness: An Essay in Genealogy,* Princeton University Press, 2002.